Ethical Sourcing in the Global Food System

This book is dedicated to our children, Lucy, Kim and Ricardo, who represent the future of ethical sourcing. It is also dedicated to the loving memory of Annabel Ware, who took a keen interest in the issues covered here.

Ethical Sourcing in the Global Food System

Edited by
Stephanie Barrientos
and
Catherine Dolan

London • Sterling, VA

First published by Earthscan in the UK and USA in 2006
Reprinted 2007

ISBN: 978-1-84407-199-9 paperback
       978-1-84407-189-0 hardback

Typesetting by JS Typesetting Ltd, Porthcawl, Mid Glamorgan
Printed and bound in the UK by Antony Rowe, Chippenham
Cover design by Yvonne Booth

For a full list of publications please contact:

Earthscan
8–12 Camden High Street
London, NW1 0JH, UK
Tel: +44 (0)20 7387 8558
Fax: +44 (0)20 7387 8998
Email: earthinfo@earthscan.co.uk
Web: www.earthscan.co.uk

22883 Quicksilver Drive, Sterling, VA 20166-2012, USA

Earthscan publishes in association with the International Institute for Environment and
Development

A catalogue record for this book is available from the British Library

Library of Congress Cataloging-in-Publication Data

Ethical sourcing in the global food system / edited by Stephanie Barrientos
and Catherine Dolan.
    p. cm.
  Includes index.
  ISBN-13: 978-1-84407-198-2 (pbk.)
  ISBN-10: 1-84407-198-7 (pbk.)
  ISBN-13: 978-1-84407-197-5 (hardback)
  1. Food industry and trade–Moral and ethical aspects. 2. Food industry
and trade–Social aspects. 3. Food industry and trade–Environmental
aspects. I. Barrientos, Stephanie. II. Dolan, Catherine.
  HD9000.5.E85 2006
  174'.9664–dc22

                                                2006001116

The paper used for the text of this book is FSC certified.
FSC (The Forest Stewardship Council) is an international network
to promote responsible management of the world's forests.

FSC
Mixed Sources
Product group from well-managed
forests and other controlled sources
Cert no. SGS-COC-2953
www.fsc.org
© 1996 Forest Stewardship Council

Printed on totally chlorine-free paper

Contents

List of Figures, Tables and Boxes

Figures

Tables

Boxes

List of Contributors

Diana Auret is an anthropologist who has spent 26 years in development work, the last eight of which have been spent in the development of social auditing. She participated in the Ethical Trading Initiative pilot in Zimbabwe, and has since trained auditors and participated in local, multi-stakeholder associations in Kenya, Zambia, Zimbabwe and South Africa. She has audited for supermarkets and local stakeholder bodies in various African countries, and has been working for FLO-Cert (Fairtrade Labelling Organizations International-Cert) as the FLO-Cert Southern African coordinator. She is particularly interested in gender and cultural issues, and has attempted to instil a better understanding of them in the training of auditors.

Stephanie Barrientos is a fellow at the Institute of Development Studies (IDS), University of Sussex, UK. She has researched and published widely on gender and development in Africa and Latin America, globalization and informal work, corporate accountability, fair trade, ethical trade and international labour standards. She coordinated a research project on Gender and Ethical Trade in African Horticulture (funded by the UK Department for International Development, or DFID) and has been engaged in a research programme on the mainstreaming of fair trade (funded by the Ford Foundation and the Leverhulme Foundation). She coordinated the UK Ethical Trading Initiative Impact Assessment (2003–2005) with studies in India, Costa Rica, Viet Nam, South Africa and the UK. She has co-authored papers and made a video on participatory social auditing with Diana Auret (co-funded by DFID and the Levi Strauss Foundation).

Henk Campher was a policy adviser: private sector at Oxfam until 2004, when he focused on the impact of the private sector on poverty and sustainable development and helped to lead the Oxfam Coffee Campaign. He is currently the director of corporate policy and practices at the International Business Leaders' Forum. He focuses on developing innovative and forward-looking corporate policy and practice solutions for business leadership and corporate engagement in sustainable development. Henk acted as special adviser to UK Prime Minister Blair's Commission for Africa and is a founding member of Business Action for Africa. He has authored a number of key business and development reports.

David Croft was head of Co-op Brand and Technical at the Co-operative Group until 2005. In that role he managed the own brand range and an international supply chain supporting over 4000 different products, and worked to develop fair trade product ranges and supply chain ethical sourcing activity in a commercial environment. David has also been a director of the Ethical Trading Initiative, a unique partnership of industry, trade unions and non-governmental organizations (NGOs) working to develop solutions on ethical trade. He is now ethical sourcing director for Cadbury Schweppes plc.

Catherine Dolan is assistant professor of anthropology at Northeastern University, US, and research fellow in the African Studies Center, Boston University. Her research focuses on the political economy of food and agriculture; gender and globalization; ethical sourcing and moral consumption practices; and the social dimensions of commodity chains in Africa. She has been engaged in two research projects on Gender and Ethical Trade in African Horticulture (funded by DFID), and is currently beginning research on the socio-economic impact of fair trade tea in Kenya (funded by the National Science Foundation). Her articles have recently appeared in *Development and Change*, *Rural Sociology*, *Environment and Planning A*, *World Development*, *Journal of Asian and African Studies* and *Journal of Consumer Culture*.

Tom Fox works for the United Nations Development Programme (UNDP) in Zambia. He leads UNDP Zambia's work on private sector development and manages the Growing Sustainable Business Initiative, which facilitates business-led enterprise solutions to poverty through supply chain linkages and other partnerships. Until April 2005 he worked for the International Institute for Environment and Development (IIED), where he carried out research, analysis and project management in the field of business and sustainable development, with a particular focus on corporate social responsibility in the South. With Bill Vorley, he also brokered the Race to the Top project, a multi-stakeholder initiative addressing the UK supermarket sector's impacts on sustainable development.

Don Pollard has been active in the Rural and Agricultural Section of the Transport and General Workers' Union for over 25 years, where he has served as a branch secretary and chaired both regional and national committees. He has presented workers' interests at national and international fora, including the issue of gangmasters in the UK, and the conditions facing workers and small-scale producers in the banana industries of Latin America and the Caribbean. He has written several reports on gangmasters in the UK, and has worked with the Ethical Trading Initiative (ETI) and government agencies to identify and recommend effective action to eliminate the abuse of workers. Don also represented the International Union of Food Workers (IUF) on the ETI Banana Pilot and was a director of Banana Link.

Marina Prieto-Carrón holds a doctorate from the University of Bristol, UK, where she is currently a post-doctoral research fellow in the Department of Politics. Her research has focused on women workers in banana plantations and textile factories in Central America (mainly Nicaragua), and specifically how corporate codes of conduct can respond to the needs and interest of women workers. She has recently engaged with the corporate social responsibility (CSR) agenda of multinationals and their supply chains in developing countries. Since 1999, she has been involved in the management committee of the Central America Women's Network (CAWN), and from 2000 to 2004 was a Research Associate of the New Academy of Business (NAB).

Laura Raynolds is an associate professor of sociology and the co-director of the Fair Trade Research Group at Colorado State University, US. Her research focuses on fair, alternative and organic markets and movements, global agro-food networks, and the restructuring of international production organization and rural labour forces. Recent articles by Laura appear in *World Development, Sociologia Ruralis, Agriculture and Human Values* and the *Journal of International Development*.

Sally Smith is a research officer at the Institute of Development Studies, University of Sussex, UK. She has an MA in Rural Development from the University of East Anglia. She is a social scientist with specialization in fair trade and corporate social responsibility. Her research interests include international trade, labour and poverty; plantation agriculture and work; and institutions for smallholder and worker empowerment. She has skills in qualitative and participatory research, gender analysis, impact assessment, and monitoring and evaluation. She has research experience in African horticulture and coffee, Asian garments and footwear, and Central American coffee and bananas.

Anne Tallontire is a specialist on fair trade and other approaches to ethical trade. She works at the Natural Resources Institute, University of Greenwich, UK, where she is part of the Enterprise, Trade and Finance Group and the Natural Resources and Ethical Trade programme. Her work focuses on socio-economic and institutional issues in ethical trade, particularly trading relationships along supply chains; the inter-relationship between standards and smaller producers and other marginalized groups; the implications of export trade for small-scale producers and agricultural workers; and gender issues in ethical trade, particularly in the coffee, forest and horticulture sectors. She teaches a course on ethical trade as part of the University of Greenwich MA in World Trade and Development. Her clients have included FLO-International, DFID, the World Bank and the Common Fund for Commodities.

Bill Vorley leads the Sustainable Markets Group at IIED and is part of IIED's Business and Sustainable Development programme. He previously worked at the Institute for Agriculture and Trade Policy in Minnesota, US, and at the Leopold Centre for Sustainable Agriculture in Iowa. His work at IIED focuses on agri-food systems, agribusiness and market structure, especially regarding the access of small-scale farmers to dynamic markets. He has an insider's knowledge of agribusiness and has written extensively on the sector's response to the challenge of sustainability. Bill is one of the coordinators of the multi-partner Regoverning Markets programme and, with Tom Fox, has also brokered a process with civil society and multiple retailers around improving the social and environmental performance of the UK supermarket sector.

Acknowledgements

This book has come about as a result of many years of research and investigation into issues concerning ethical sourcing. Our focus has always been on more marginalized producers and workers. They provide the backbone to food production, but often struggle to maintain a living within the global food system. These are the groups who are often overlooked by mainstream commercial operators, and in conventional academic and policy analysis. The pressure for ethical sourcing has provided an important opportunity to address their situation and to improve their lives. This book is a tribute to all of those producers and workers who have participated over the years in research projects that are reported in this collection.

Many people have contributed their time and support to the writing of this book. We would like to thank all of those who have participated in the discussions, seminars and workshops on fair trade and ethical trade that have underpinned the writing of many of the chapters in this volume. We would also like to thank Sally Smith, Armando Barrientos and Mick Blowfield for reading and commenting on some chapters, and for their patience and support during the writing process. In addition, we would like to thank Lindsey Napier for her help preparing the Glossary. Nadine Beard provided essential assistance and coordination throughout the editing of this book, which would not have come to fruition without her.

March, 2006
Brighton

Preface

Tim Lang

After years of being marginal in debates about both consumerism and capitalism, the ethical dimension is back, nowhere with more presence than in food matters. It could be argued that the question of ethics in food sourcing never really went away, and was only submerged rather than annihilated. Nonetheless, we should celebrate – even while analysing why and where – that morality is once more being woven into food policy discourse.

To some extent, the subject of this book (and the many other books, reports and articles exploring and promoting the issue) is a testament to the persistence of civil society organizations. Surely it is they who developed the methods, the arguments and the constituencies that demanded better tracking of just how their consumer goods – in this case, food – got to them. Even as we immerse ourselves in the complexities covered in this book, and relish how governments and the food supply chain have to engage with the problems of delivery, we should remember the work, often in the teeth of vehement and sometimes bloody mainstream opposition, of those who carved out the ethical sourcing tracks before us.

We should remember the work decades ago by, for example, the anti-apartheid movement asking us to boycott South African foods; the campaign to pressurize Nestlé, then as now one of the world's giant food companies, to refrain from inappropriate sales of breast milk substitutes and to obey United Nations marketing codes of practice; and the demands of migrant Californian farm workers that we should not enjoy US grapes if they came with the blood of exploited labour as a hidden ingredient. All of these campaigns, and many smaller, less high-profile ones, carved out the space for us to explore the complexities mapped in this book.

To some extent, the fact that we even tussle with ethical sourcing – what it is, whether it can be delivered, how to monitor it, whether it can be trusted, how it fits other demands for information – is a testament to what Yiannis Gabriel and I in *The Unmanageable Consumer* (Gabriel and Lang, 2006) call the new wave consumer groups. These emerged, alongside alternative trade organizations (ATOs), from the post-World War II Western consumer revolt. They result, in part, out of disdain for materialist excess, in part from a desire to put quality over quantity, and in part to claim difference – the inevitable and welcome desire of new generations to pioneer a better way of doing things.

In truth, ethical sourcing as a significant force has its roots in the early 19th century. No sooner was industrial capitalism in full swing than a desire for another vision for progress emerged. In the UK, this found expression in the labour, socialist, anti-slavery, Methodist and, not least, co-operative movements. These and other movements staked a claim on progress that put ethical sourcing at their heart. Daily living meant choices not just about what to do, but how. The Co-operative Movement, in particular, gave voice to that era's definition of ethical sourcing: food should be unadulterated, affordable, decent and from sources under the consumers' collective control. That collectivity was essential; today's individual angst about consumer choice would seem bizarre to a true mid 19th-century co-operator.

That food should be such a significant battleground for refining both the product and the process is not surprising. In Tony Winson's (1993) memorable phrase, food is an 'intimate commodity'. When it enters our bodies, our identities are shaped through it and we take in an entire and sometimes poorly understood appreciation of the supply chain. Some contemporary industry analysts argue that consumers neither want nor need to understand the complexities behind the check-out till. The brand is the seal of trust. But when we eat, we partake in an increasingly long chain of reactions. If brand is the sole mediator of trust, this is a fragile relationship.

In truth, few of us choose as much as we think we do. We select rather than choose. Our much-vaunted choices are framed by context, class, culture, family, history and the panoply of social determinants, not least the vast marketing expenditure pitched at building brand loyalty. Yet, we all learn, whatever our circumstances or cultures, to distinguish between freedoms and constraints; some have more, others far less. It is this aspiration that the current debate about ethical sourcing taps into. The demand for things to be better – to reassert control – has grown, fed, too, by the public unease and sometimes horror at exposés of adulteration, exploitation, fraud, unsound production and farming, and poor nutrition during recent years.

I see no end to this debate. Nor should we wish it to end. In a culture that is so heavily framed by the quick fix, the throw-away instant celebrity, short-termism and me-too-ism, the re-emergence of a solid, articulate challenge about ethics and process is frankly wonderful. We are engaged in nothing less than trying to reassert the importance of human decency and our future in a food culture under strain to go in a different direction. The movement for ethical sourcing is already helping to articulate the need for new rules for daily living. These are matters not for you or me as individuals, but for us all. Our right to the term civilized hangs on issues such as these.

Tim Lang
Professor of Food Policy, City University, London
March 2006

References

Gabriel, Y. and Lang, T. (2006) *The Unmanageable Consumer*, 2nd edition, Sage, London

Winson, A. (1993) *The Intimate Commodity*, Garamond Press, Aurora, Ontario

List of Acronyms and Abbreviations

ABF	Associated British Foods
AEAAZ	Agricultural Ethics Assurance Association of Zimbabwe
ALP	Association of Labour Providers (UK)
ANPED	Northern Alliance for Sustainability
ASPROLPA	Association of Milk Producers in the Province of Aroma (Brazil)
ATC	Associación de Trabajadores del Campo (Nicaragua)
ATO	alternative trade organization
AWB	Agricultural Wages Board (UK)
AWO	Agricultural Wages Order (UK)
AWT	Agricultural Wages Team (UK)
BAFTS	British Association for Fair Trade Shops
BANANIC	Nicaraguan Banana Producers' Association
BBC	British Broadcasting Corporation
BRC	British Retail Consortium
BSE	bovine spongiform encephalopathy
C	Celsius
CAWN	Central America Women's Network
CCCC	Common Code of Conduct for Coffee
CEO	chief executive officer
CEP	Council on Economic Priorities
CEPAA	Council on Economic Priorities Accreditation Agency
CJD	Creutzfeldt–Jakob disease
COBIGUA	Compañia Bananera Independiente de Guatemala SA (Independent Banana Corporation of Guatemala)
COLE–ACP	Liaison Committee Europe–Africa, Caribbean, Pacific
COLSIBA	Latin American Banana Workers' Union Coordination
COVERCO	Commission for the Verification of Codes of Conduct
CR	corporate responsibility
CSR	corporate social responsibility
DEFRA	UK Department for Environment, Food and Rural Affairs
DFID	UK Department for International Development
DTI	UK Department of Trade and Industry
ECOSOC	United Nations Economic and Social Council
EFA	European Federation of Agricultural Workers

EFTA	European Fair Trade Association
ESRC	Economic and Social Research Council
ETI	Ethical Trading Initiative (UK)
EU	European Union
EurepGAP	Euro Retailer Producer Working Group Standard for Good Agricultural Practice
FTF	Fair Trade Federation
GLACC	Global Alliance on Coffee and Commodities
GM	genetically modified
GMIES	Salvadorian Independent Monitoring Group
GRI	Global Reporting Initiative
FES	Friedrich-Ebert-Stiftung
FGD	focus group discussion
FIAN	FoodFirst Information and Action Network
FINE	forum of *FLO*, *IFAT*, *NEWS* and *EFTA*
FLO	Fairtrade Labelling Organizations International
FPC	Fresh Produce Consortium (UK)
FPEAK	Fresh Produce Exporters Association of Kenya
FTO	Fair Trade Organization
HACCP	Hazard Analysis Critical Control Points
HEBI	Horticultural Ethical Business Initiative (Kenya)
HPC	Horticultural Promotion Council
H&S	health and safety
ICO	International Coffee Organization
IDS	Institute of Development Studies (UK)
IFA	International Framework Agreement
IFAP	International Federation of Agricultural Producers
IFAT	International Fair Trade Association (*formerly* International Federation of Alternative Trade)
IFOAM	International Federation of Organic Agriculture Movements
IIED	International Institute for Environment and Development
ILO	International Labour Organization
IPEC	International Programme on the Elimination of Child Labour
IRFT	International Resources for Fairer Trade
ISEAL	International Social and Environmental Accreditation Labelling
ISO	International Organization for Standardization
IUF	International Union of Food, Agricultural, Hotel, Restaurant, Catering, Tobacco and Allied Workers' Associations
KFC	Kenya Flower Council
kg	kilogram
km	kilometre
MAFF	UK Ministry of Agriculture
MEP	Member of the European Parliament

MERCOSUR	*Mercado Común del Sur* – the Southern Common Market
MORI	Market and Opinion Research International (UK)
MP	Member of Parliament (UK)
MSF	Médecins Sans Frontières
NAB	New Academy of Business
NEWS!	Network of European Worldshops
NFU	National Farmers' Union
NGO	non-governmental organization
NMW	National Minimum Wage (UK)
NRET	Natural Resources and Ethical Trade Programme
NUAAW	National Union of Allied and Agricultural Workers
NZTT	National Resource Development College/Zambia Export Growers Association Training Trust
OCA	Organic Consumers' Association (US)
OECD	Organisation for Economic Co-operation and Development
ORSEU	Office for European Social Research
PIL	public milk company (Brazil)
PLA	participatory learning and action
PRA	participatory rural appraisal
PSA	participatory social auditing
PSN	Producer Support Network (of FLO)
PwC	Price Waterhouse Cooper
RHM	Rank Hovis McDougall
SA	Social Accountability
SAI	Social Accountability International
SAN	Sustainable Agriculture Network (UK)
SAP	Structural Adjustment Policy
SARS	severe acute respiratory syndrome
SASA	Social Accountability in Sustainable Agriculture
SAWS	Seasonal Agricultural Workers Scheme
SCF	Save the Children Fund
SEDEX	Supplier Ethical Data Exchange
SFEO	small farmer economic organization
SSC	Stakeholders Steering Committee (of HEBI)
T&GWU	Transport and General Workers' Union (UK)
TLWG	Temporary Labour (Gangmaster) Working Group (of ETI)
TRABANIC	Nicaraguan Banana Workers' Association
TUC	Trade Union Congress
UK	United Kingdom
UNCTAD	United Nations Conference on Trade and Development
UNCTC	United Nations Centre for Transnational Corporations
UNDP	United Nations Development Programme
UNHDR	United Nations Universal Declaration of Human Rights
UNSITRAGUA	Unión Sindical de Trabajadores de Guatemala

US	United States
USFT	United Students for Fair Trade (US)
VAT	value-added tax
WDM	World Development Movement
WIETA	Wine Industry Ethical Trade Association (South Africa)
WTO	World Trade Organization
ZEGA	Zambia Export Growers Association

1

Transformation of Global Food: Opportunities and Challenges for Fair and Ethical Trade

Stephanie Barrientos and Catherine Dolan

Introduction

Over the past decade a paradox has emerged in the global food system. On the one hand, we have seen a rapid rise in corporate domination of food production, particularly, food retailing. Consumers have become increasingly dependent upon large supermarket chains for year-round food, creating fierce competition and aggressive commercial practices amongst leading corporate players. On the other hand, consumers have also become more discerning about the social impacts of food sourcing. Many civil society organizations (non-governmental organizations, or NGOs, and trade unions) have criticized the adverse social effects of corporate power on small producers and workers in global agriculture. Companies are increasingly under pressure to enhance the position of small producers and workers in their supply chains. This has resulted in the rise of a range of voluntary initiatives under the umbrella of ethical sourcing, including fair and ethical trade.

There are now a plethora of voluntary initiatives and organizations linked to fair trade and ethical trade operating in different parts of the world (see the Glossary to this volume for a summary of examples). They aim to ensure that the small producers and workers linked to the global food system benefit from or, at the very least, are not harmed by company sourcing strategies. Some are individual company schemes to improve conditions for small producers and workers in their own supply chains – such as the 'preferred supplier' scheme of Starbucks and supermarket codes of labour practice. Some are wider sector-based initiatives involving groups of companies and other stakeholders – such as the International Cocoa Initiative. Others are independent initiatives that cross sectors and involve different stakeholders, such as the Fairtrade Foundation and the Ethical Trading Initiative (ETI). Some rest on reputation while others operate labelling schemes. It can be difficult for a discerning consumer to navigate this diverse landscape. Increasingly, consumers are confused by the barrage of slogans, labels and schemes, and are uncertain whether or what benefits accrue to the stated beneficiaries in a complex global food system.

The aim of the book is to unravel some of the main ethical sourcing initiatives that have arisen in the rapidly changing context of global food production and retailing. The rise of ethical sourcing raises important questions. What factors are driving the growth of fair and ethical trade? Are civil society organizations promoting ethical and fair trade genuinely able to affect the commercial practices of large corporate producers and retailers? Can a global food system driven by commercial pressures seriously incorporate ethical sourcing? What role does the monitoring and enforcement of social codes play in advancing their credibility? Are small producers and workers who are linked to the global food system genuinely able to benefit from this trend? These are the questions that will be explored through the different perspectives and case studies in this book.

Background

The increase in ethical sourcing is directly linked to changes that have taken place throughout the global food system, from production to consumption, during the last few decades. The past decade, in particular, has seen the rapid rise in supermarket retailing, not only in Northern but increasingly in Southern countries (Reardon and Berdegué, 2002; Fox and Vorley, Chapter 10, this volume). Supermarket distribution chains have become streamlined as they move away from the use of traditional food markets to centralized procurement systems and more direct year sourcing from suppliers. This is often accompanied by vertical integration or relational contracting, where production, processing, marketing and financing are linked. Large food manufacturers and distributors, such as Unilever and Nestlé, continue to play an important role in food sourcing; but their position has been eclipsed by supermarkets at the consumer end of the chain. The unrivalled market power of global food manufacturers and retailers has meant that suppliers are ever more dependent upon them for market access. Increasingly, staying in business hinges on meeting the stringent terms and conditions for price, quality and delivery set by large corporate buyers. The food system has thus become more integrated and expansive in its global reach, but increasingly driven and defined by large corporate players (Vorley, 2004).

Changes in the global food system have brought benefits to some, particularly more affluent consumers in the North, but also amongst some sections of the population in the South. As these consumers come to expect more from the global food system, food industries have responded with new products, packaging and innovations to match shifting consumer tastes. Consumers are now able to purchase a wide range of basic and 'exotic' food from different parts of the world throughout the year. As more women engage in paid work, supermarkets provide a convenient means of reducing the time spent on shopping, and the availability of processed and prepared foods reduces the time spent on cooking and food preparation in the household.

Consumer concerns about the nature of food production and its health and safety implications have also grown. While European consumers have long experienced anxieties regarding food quality and safety, a series of 'food scares' and safety scandals erupted during the 1980s to 1990s (e.g. outbreaks of BSE, CJD,[1] listeria, salmonella and foot and mouth disease) that dramatically increased concerns surrounding food products. Questions have also been raised about the environmental, health and sustainability impacts of the global food system. Environmentalists have pointed to the risks to health and environment from mono-cultivation, the use of pesticides and more intensive farming methods in agriculture. They criticize the polluting effects of 'food miles' involved in the transportation of food from ever-more distant parts of the globe to the plates of Northern consumers. Health experts have pointed to the problems of rising obesity and diet-related diseases stemming from our growing reliance on highly processed foods that are low in nutritional content. Activists are concerned that export-oriented development has triggered a shift away from local food production, undermining food security in developing countries and threatening the livelihoods of small farmers.[2]

More recently, a growing lobby of civil society groups has also questioned the social impacts of modern global food production, which is the main concern of this book.[3] This has been facilitated by heightened exposure to global issues through the media, internet and long-haul international travel. It has also been boosted by the campaigns of some civil society organizations, which have widely publicized the adverse effects of global food production on smaller producers and workers. There is, thus, a paradox in the contemporary global food system: many consumers who are capitalizing on the supermarket economy through faster, cheaper and more convenient food have become more concerned about the social conditions under which their food is produced and distributed.

The rise of ethical sourcing

As the global food system has evolved, civil society organizations, particularly NGOs, consumer groups and trade unions, have become increasingly active in advocacy and campaigning for more ethical principles in global food sourcing. Their focus is particularly on small producers and waged workers whose production and employment conditions are directly or indirectly affected by the global restructuring of agro-food markets and the trading practices of large corporate buyers. While fair trade and ethical trade initiatives share a common ideological commitment to social development, they embrace different aims and methods. Fair trade has its roots in the solidarity and charity movements of the mid 20th century and largely focuses on providing support for small producers marginalized by the global trading system. Ethical trade based on codes of labour practice came to prominence during the mid 1990s as part of a renewed emphasis on corporate social responsibility and self-regulation in the context of privatization and government deregulation. The better of them embody core International Labour Organization (ILO) conventions covering freedom of association, no child or forced labour and no discrimination. They complement a number of wider multilateral initiatives, such as the ILO Tripartite Declaration of Principles Concerning Multinational Enterprises and Social Policy, the Organisation of Economic Co-operation and Development (OECD) Guidelines for Multinational Enterprises, and the United Nations Global Compact. All provide voluntary approaches to enhancing labour standards in global production systems where key corporate buyers are often beyond the control of national regulation in the countries from which they source.

Civil society organizations (NGOs and trade unions) have long been concerned about the adverse effects of globalization on both small producers and workers. In many commodities, small producers are facing falling international prices, even to below their costs of production, often pushing them into crippling debt. They are finding it increasingly difficult to meet the quality standards set by large corporate buyers. Many are being squeezed out of global supply chains as large supermarkets concentrate their supply base (see Chapter 10). At the same time, large production units have generated a growing agricultural labour force in many areas linked to global food production. Much of this labour is female and often draws on migrant workers. Employment in agriculture and food production is largely temporary and insecure, often with poor working conditions, long hours and lack of legal or social benefits (see Chapter 5).

Civil society organizations have attempted to address these concerns through different ethical sourcing strategies. Early movements focused on creating alternative trading channels for poor producers so that they could access Northern markets without going through mainstream commercial food chains and reach socially conscious consumers. Fair trade was launched to extend these channels into more mainstream retailing, and to reach a wider range of consumers prepared to pay more for food that was certified as having been produced and distributed under more equitable conditions (Barratt-Brown, 1993).

Civil society organizations have also been able to take advantage of increased centralization within the global food system to pressure large corporate buyers and retailers. Where food is produced and distributed via open markets, it is difficult to trace its origins to a specific retailer, manufacturer or importer, let alone relate any malfeasance to a particular production site. Fragmented supply chains conceal the social relations and exploitative practices of production. By contrast, where food is produced in more integrated supply chains, it is possible to trace the effects of production on specific groups of small producers or workers, and link any adverse impacts to specific manufacturers or supermarkets. This has rendered large corporations, who occupy the most visible and powerful position in these supply chains, more vulnerable to negative publicity, linking them to the poor conditions of their overseas suppliers even where they are not formally responsible through direct ownership or employment. Such publicity can tarnish corporate images and potentially affect market share (particularly amongst the more affluent section of the consumer market) and company share prices.

However, some large retailers and producers within the food industry also want to actively project a positive corporate image. They appeal to consumers on the basis that they take responsibility for all aspects of the food they purchase, including the quality and hygiene of the food, as well as the social and environmental impacts of its production. They aim to encourage and maintain consumer loyalty in a competitive commercial environment in part, on the basis of the ethical principles and values represented by the company itself. Promoting fair and ethical trade can play an important part in sustaining this ethos and in differentiating some companies from others within the food industry.

While there are a plethora of ethical sourcing initiatives aimed at targeting the market of an ethically minded public (as summarized in the Glossary to this volume), the diversity of initiatives and ambiguity of terminology can be bewildering. A myriad of products, from fair trade milk chocolate, to shade-grown coffee, to ethical investments are now presented as part of 'alternative' consumption practices that encourage social justice and fair trade (Johnston, 2001). Yet, what is the difference between eating a banana that is fairly traded from one that is ethically sourced? Does ethical sourcing mean paying a farmer a fair price or ensuring that workers enjoy decent working conditions or both? How do fair trade and fair trade labelling differ? The answers to such questions are not always clear cut. The following sections first define and then examine the main aspects of two of the most prominent ethical sourcing initiatives in the global food industry: fair trade and ethical trade. Finally, we compare the two and weigh up the challenges facing them in the context of the wider global trading system.

Definitions and scope of ethical sourcing

Ethical sourcing covers a wide number of voluntary initiatives concerned with the ethics of production and trade. Given the plethora of initiatives, it is important to clarify the terminology we use, which can be confusing since there is no common agreement. In this book we focus primarily on the social dimensions of ethical sourcing – fair and ethical trade.[4] Fair trade focuses on equity in trading relations, and particularly support for small producers and farmers. Ethical trade covers employment conditions of workers through the implementation of codes of labour practice in the supply chains of large food corporations and retailers.

'Fair trade', at its most inclusive, is both a movement and a set of business initiatives that embodies an alternative approach to conventional international trade (OPM/IIED, 2000). The term 'fair trade' refers to the broad concept of more equitable trading relations for producers and workers, as used by the fair trade movement. This is distinguished from the 'Fairtrade' (one word) mark, which refers to the specific labelling system controlled by Fairtrade Labelling Organizations International (FLO) and its member organizations. While definitions of fair trade vary, one of the most widely agreed definitions is that of the umbrella organization FINE,[5] which has adopted the following:

> Fair trade is a trading partnership, based on dialogue, transparency and respect that seeks greater equity in international trade. It contributes to sustainable development by offering better trading conditions to, and securing the rights of, marginalized producers and workers – especially in the South. Fair trade organizations (backed by consumers) are engaged actively in supporting pro- ducers, awareness raising and in campaigning for changes in the rules and practice of conventional international trade.[6] (EFTA, nd)

'Ethical trade' is used to refer to codes of labour practice and voluntary initiatives aimed specifically at improving employment conditions and complying with labour standards in food supply chains. When employing the term 'code of labour practice', we use the following definition:

> In the context of ethical trading a ... code of labour practice is a set of standards concerning labour practices adopted by a company and meant to apply inter- nationally and, in particular, to the labour practices of its suppliers and subcontractors. (ETI, 2003)

Codes of labour practice aim to ensure minimum labour standards for workers by establishing guidelines on issues such as child labour, forced labour, working hours, discrimination, freedom of association, wages, health and safety, and other workplace-related issues. The most comprehensive of these codes are based on core ILO conventions[7] and the United Nations Universal Declaration of Human Rights (UNHDR).

Fair trade focuses on enhancing equity and providing niche markets for producers and their workers within trade. Codes of practice focus on providing enhanced labour conditions and on the rights of workers employed by conventional suppliers in mainstream supply chains. Fair trade schemes seek to establish economic justice by altering the principles of commercial exchange and terms of trade, whereas codes of labour practice aim to ensure minimum labour standards within an existing model of trade (Hughes, 2004). Fair trade only relates to products sourced from developing countries, although they may be sold in developed or developing countries. Codes of labour practice are applied to producers in both the North and the South, as corporate retailers and multinational companies have realized that labour issues are of concern in all countries, including their own (see Chapter 7).

The promotion of both fair trade and ethical trade has become a concern for many European supermarkets, particularly in the UK and Switzerland. While sales of fair trade food are growing rapidly in the US, codes of labour practice have been of less concern to supermarkets than apparel firms within North America. In general, supermarkets in the US stress food safety while European supermarkets tend to emphasize workers' welfare and environmental protection, as well as food safety (FAO, 2004). Hence, workers engaged in global food production for European markets are more likely to be covered by codes of labour practice than those supplying the US food sector.

Fair trade and Fairtrade labelling

Today over 1 million small-scale producers and workers are organized into fair trade schemes in over 50 countries in the South. Their products are sold in thousands of worldshops and fair trade shops, and are increasingly mainstreamed into supermarkets and other conventional retail outlets (Kocken, 2003). In 2004 alone, sales of labelled fair trade products generated an additional US$100 million for producers and workers in developing countries, due to global retail sales topping US$1 billion (FLO, 2005).

Fair trade seeks to foster sustainable development through 'trade not aid', aiming to move disadvantaged producers out of poverty by providing them with a fair return for their work and decent working and living conditions (see Box 1.1) (Raynolds, 2002). As Tallontire (see Chapter 2) notes, while all fair trade initiatives challenge the inequities of global trade, they embody different approaches to realizing this end.

The two main approaches – fair/alternative trade organizations (ATOs) and Fairtrade labelling initiatives – diverge in both their commercial orientation, as well as in their broader organizational philosophies and histories. The origins of ATOs lie in the charity and humanitarian activities of religious communities and development agencies during the mid 20th century. In the US, Mennonite and Brethren missions of the 1940s and 1950s assisted poverty-stricken communities in the South by selling their handicrafts to Northern markets (Grimes and Milgram, 2000). Similarly, in the UK and Europe, aid organizations engaged in handicrafts importation as a way of supporting marginalized producers and workers (Barratt-Brown, 1993; Ransom, 2001). By the 1970s, myriad church and development organizations had emulated this approach, establishing alternative trade links with groups and co-operatives across a range of developing countries. Oxfam is generally credited with pioneering the contemporary ATO movement through the establishment of its partnership-based Bridge Programme in 1964, which marketed handicrafts from the South as a means of economic development, offering higher returns to producers through direct trade, training and advice, and fair prices.[8] Within the last few decades, alternative trade organizations such as Equal Exchange, Bridgehead, SERRV International (formerly known as Sales Exchange Refugee Rehabilitation and Vocation), Twin Trading, Traidcraft and Ten Thousand Villages have mushroomed.[9] While individual ATOs have different priorities and emphasis, all share a common purpose to foster a trading model that respects Southern producers as equal partners in a business relationship.[10]

Box 1.1 Basic criteria of fair trade

The Fairtrade Foundation, Oxfam and Traidcraft have agreed the following basic criteria of fair trade:

- A clear set of criteria defining the fair trade terms is available to consumers and producers.
- An organization (auditor, body of trustees), independent of business interests, oversees the implementation of the fair trade principles.
- The suppliers are selected on the basis of being poor and relatively disadvantaged by the way in which the commercial market operates.
- There are monitoring systems to ensure that the fair trade principles and criteria are met and that individual producers are benefiting from the trading terms applied.
- Producers are consulted and are able to contribute to the development of the monitoring systems.
- Trading terms are mutually agreed and always give greater support to the producer than they could expect from the commercial market.

Source: www.fairtrade.org.uk/about_partnership.htm

Fairtrade labelling emerged during the late 1980s. There was growing interest in diversifying the range of fair trade products and in extending their market coverage beyond small charity shops. This effort was spearheaded in 1988 by the Dutch organization Max Havelaar, which sought to link impoverished coffee growers in Mexico with international markets through a fair trade labelling scheme that targeted mainstream retail outlets. The Max Havelaar label met with immediate success, establishing fair trade labelling as a viable marketing concept (Levi and Linton, 2003). By the 1990s, three fair trade labels – Max Havelaar, Fairtrade Mark and TransFair – were established in Europe to certify coffee, with variations of the model extended to the US and Japan. These labels have been extremely effective in boosting consumer recognition and market growth: in The Netherlands, for example, 87 per cent of the population recognized the Max Havelaar label in 1997 (ANPED, 1999), and fair trade coffee's share of the market increased 20-fold following the label's introduction into mainstream supermarkets (Conroy, 2001).

Yet, despite its success, the proliferation of schemes and fair trade labels across countries created confusion in the minds of retailers, importers and producers. Fearing that the existence of multiple labels targeted to the same ethical consumers could undermine fair trade, Fairtrade Labelling Organization International (FLO) was created in 1997 to harmonize the standards and activities of labelling initiatives (e.g. the UK Fairtrade Foundation and TransFair in the US and Canada are national initiatives of FLO). FLO is now responsible for setting international standards and for certifying and labelling fair trade products (see Table 1.1).[11] Among consumers, the Fairtrade mark offers assurances that the products they buy meet certain social and environmental standards. It ensures that the producer receives a fair price (i.e. Fairtrade banana producers receive US$1.75 per box above the world market price), and a premium is paid to invest in producer organizations and/or community improvements in healthcare, education, housing and local infrastructure (see Chapter 2; Raynolds, 2000). As of November 2005, FLO has certified 548 producer organizations representing over 1 million farmers and workers in over 50 countries in Africa, Asia and Latin America to supply fair trade products (FLO, 2006).

The rapid expansion of fair trade labelled products since 2000 has been impressive. Between 2002 and 2003, fair trade labelled sales registered remarkable growth, increasing by 42.3 per cent internationally (Fairtrade Foundation, 2004a). In the US fair trade certified products – including coffee, tea, cocoa, rice, sugar and fruit – are offered in more than 30,000 cafés, restaurants, supermarkets and dining service venues nationwide, with over 500 companies throughout the US currently licensed to sell fair trade certified products (Transfair, 2005a). This growth positions the US as potentially the most significant national market in coming years. To date the European market has witnessed the most impressive sales of fair trade labelled products, with an approximate retail value of €597 million in 2005 (UK£413) (Krier, 2005). In Switzerland, for example, 47 per cent of bananas, 28 per cent of flowers and 9 per cent of all sugar sold is now fair trade labelled. Market penetration has been particularly strong in the UK, where sales of fair trade products reached

Table 1.1 *Food products carrying the Fairtrade mark*

Cocoa	Fresh vegetables
Coffee	Green beans
Dried fruit	New potatoes
Fresh fruit	Sweet peppers
Apples	Fruit juices
Avocadoes	Honey
Bananas	Nuts and seeds
Citrus fruit	Preserves and spreads
Coconut	Rice
Grape	Sugar
Oranges	Tea
Mangoes	Wine and beer
Pears	
Plums	
Pineapples	

Source: Fairtrade Foundation, 2006

£140.00 (US$250m) in 2004, a 450 per cent increase over 1998. Fair trade labelled ground coffee has achieved a 20 per cent market share in the UK, with strong growth in bananas and tea (Fairtrade Foundation, 2004a, 2005).[12]

Much of this growth was accounted for by the expansion of fair trade goods in mainstream retail outlets such as supermarkets. Today fair trade labelled products can be purchased in over 55,000 supermarkets throughout Europe (Krier, 2005). This expansion has been particularly marked in the UK where supermarkets have registered consistent year-on-year sales. Between 2002 and 2003, Co-op sales increased by 112 per cent, Tesco by 70 per cent and Waitrose by 24 per cent.[13] Supermarket sales have been significantly boosted by the expansion of supermarket 'own-brand' fair trade-labelled products.[14] The Co-operative Group in the UK spearheaded this trend in 2000 with the launch of its own-brand chocolate bar (with Divine) (see Chapter 4). Sainsbury's own-brand fair trade label appeared on the shelves in October 2001. In 2004, Tesco also launched its own-brand fair trade line and now carries a Tesco fair trade range of tea, coffee, chocolate, cookies and orange juice, as well as fruit including bananas, mangos and plums (Moore et al, 2004; Smith and Barrientos, 2005). In addition to supermarket sales, some large food producers and distributors (e.g. Nestlé) are also developing fair trade lines, with others expected to follow (Nicholls and Opal, 2005).

As Tallontire (see Chapter 2) illustrates, while both FLO International and ATOs share a common vision of trade as a way to promote development and reduce poverty, there are some differences in their strategies. Fair trade labelling organizations are not involved in commercial exchanges themselves but issue (and monitor) fair trade marks to manufacturers or importers that subsequently sell fair trade-labelled products in commercial outlets (see the Glossary to this volume). ATOs act as importers and depend upon their brand reputation and name recognition to communicate their values (Young, 2003). They have conventionally marketed their products through traditional fair trade outlets such as world shops, NGO charity shops and specialist mail-order companies, many of which are not typically marked with an independent verification label. This is changing, however. ATOs have launched several mainstream products, including high-profile brands such as Cafédirect and Divine Chocolate, and are increasingly selling through supermarket chains. In 2004, the International Fair Trade Association (IFAT) launched the Fair Trade Organization (FTO) mark. While not a product label, the FTO mark identifies fair trade organizations, enabling them to differentiate themselves from other commercial organizations engaged in fair trade.

Fair trade labelling organizations have, until recently, focused on the labelling of food products. By far the largest market for fair trade food is coffee, which forms the core of labelling initiatives in Europe and North America, although the market for tea, chocolate and bananas is expanding. Recent reports peg the total European market for fair trade coffee at over US$300 million, produced by 550,000 farmers in 300 organizations worldwide (Raynolds, 2002; Traidcraft, 2003). In 2003, UK-based Cafédirect alone returned UK£2.8 million (US$5 million) to its coffee suppliers (Nicholls and Opal, 2005). While fair trade coffee currently represents only a small share of the US market (1.8 per cent in 2004), retail sales reached close to US$370 million in 2004 (TransFair, 2005b), with TransFair US estimating a return of US$30 million (UK£16.8 million) to coffee producers in developing countries over a five-year period (Nicholls and Opal, 2005). As labelled coffee becomes more widely available in US supermarkets (e.g. Costco, Sam's Club, Stop & Shop, Giant, Tops, Fred Meyer and Albertsons), as well as in restaurants such as McDonald's,[15] Dunkin Donuts and Starbucks, the upward trend is likely to accelerate.

Tallontire (see Chapter 2) concludes that despite some differences between ATOs and labelling organizations, both initiatives are committed to increasing consumer awareness and expanding into mainstream markets. Fair trade products are now stocked in many European supermarket outlets. In the US, where the retail sector is more diverse, 7000 retail outlets now stock fair trade coffee, including natural food chains, café chains such as Starbucks and even convenience stores (Conroy, 2001). However, while the retail penetration of fair trade products continues to expand in both North America and Europe, Tallontire argues that there remains a salient tension between the commercial and development objectives of the two approaches, leading to significant differences in the way that ATOs and Fairtrade labelling initiatives manage and monitor their trading relations. These tensions are likely to deepen as fair trade lines are increasingly adopted by large European retailers and food manufacturers.

The capacity of ethical sourcing to create more equitable and sustainable production is also explored by Raynolds (see Chapter 3), who compares the evolution of fair trade with that of the organic movements in the US. The aims of the two movements differ: fair trade centres on the social conditions of production, whereas organic certification focuses on the ecological conditions of production. However, there are signs that the two movements are forging a common ground in defining minimum social and environmental requirements. Yet, Raynolds suggests that fair trade's specific focus on the inequalities in the global trading system renders it a more effective oppositional movement. By revealing the true nature of global relations of exchange and challenging market competitiveness based solely on price, Raynolds asserts that fair trade can transfer greater control of the agro-food system to producers in developing countries and help to bridge the North–South divide. The empowerment that results from this process is not simply a function of increased incomes, but rests on long-term relationships, direct trade and credit provision, all of which can help small producers to eventually expand into non-fair trade markets (Nicholls and Opal, 2005).

Ethical trade and codes of labour practice

The rise in ethical sourcing has also been marked by a rapid expansion in codes of labour practice and multi-stakeholder initiatives around labour standards. The rapid proliferation of codes is linked to changes that have arisen in a globalized economy, including deregulation, technological innovation and trade liberalization, which have dramatically altered the way in which companies and governments operate. Governments have conventionally been assigned the role of ensuring a stable framework for statutory labour standards and employment stability. However, globalization has led to intensified international competition between countries to attract foreign direct investment and export orders based on low costs. Many governments now offer tax havens and have designated export processing zones in which much labour regulation is relaxed (Enloe, 1990). At the same time, deregulation has reduced the capacity of states to regulate global production and has devolved more responsibility for social and economic welfare to the private sector. The renewed focus on international labour standards and the rise of voluntary codes of labour practice are, in part, a response to these tendencies as civil society organizations work to prevent a 'race to the bottom' in which the depression of wages and labour conditions becomes a competitive advantage in the global economy (Sengenberger, 2002).

At the same time, globalization has engendered unparalleled advances in global communications and information technology that have allowed for closer scrutiny of corporate behaviour and provided new opportunities for global coalitions of advocacy groups to exert pressures on corporations. The food industry has been a prime target of NGO campaigns (Christian Aid, 1997). Media exposés of corporate wrongdoing, ranging from worker abuse in the Chilean grape and Kenyan vegetable industries, to migrant workers enslaved by UK gangmasters, to Californian tomato pickers exploited by Taco Bell, are now ferried across the internet in seconds and circulated in a range of media fora (Christian Aid, 1996).

The transformation of the global food system has increased the vulnerability of food companies and retailers to such exposure. The risks faced by supermarkets, in particular, are exacerbated by unprecedented retail concentration. In the UK, for example, according to market analysts in 2005, five supermarket chains accounted for 74 per cent of the food retail market, with one company, Tesco, accounting for 30 per cent of the market (Taylor Nelson Sofres, 2005). Direct sourcing, with suppliers packing products in supermarket-branded packaging at source, make it easier for civil society organizations to identify and expose adverse employment practices in the supply chains of named supermarkets. In a commercial environment where reputation and brand integrity are evermore critical to profit margins, bad publicity can have significant effects on company value. This is particularly the case where tight margins mean that even a small loss of market share can have devastating consequences for a company. In this climate, applying codes of labour practice not only addresses the concerns of socially conscious consumers (i.e. ethics pays), but serves as a deterrent against brand-damaging campaigns.

Company codes

Voluntary company codes began to proliferate in the food system during the early 1990s. A number of Northern companies developed in-house company codes that established guidelines for product characteristics (size, quality, specifications, etc.) and the technical aspects of production (food safety, pesticide issues and integrated crop management) (Blowfield, 1999). These partly resulted from the series of food scares that occurred in the UK during the 1980s and 1990s, and prompted a legislative reformulation of food standards. With the transfer of the legal responsibility for ensuring food safety to the private sector, traceability from farm to fork has become the norm in the European food industry. A number of voluntary standards were also introduced by supermarkets, including the Euro Retailer Producer Working Group Standard for Good Agricultural Practice (EurepGAP), covering food hygiene, safety and good agricultural practice, and the British Retail Consortium (BRC) Technical Standard, which establishes a food safety and monitoring protocol for retailer-branded food products. Both EurepGAP and BRC incorporate the principles of Hazard Analysis Critical Control Points (HACCP), an approach to food safety that is now a requirement among many European supermarkets (Dolan and Humphrey, 2004). Codes of labour practice constitute an additional form of voluntary regulation that suppliers must comply with to sell to leading food companies and retailers.

Following the Earth Summit in Rio de Janeiro in 1992 and the anti-sweatshop campaigns in the garment sector, food retailers and distributors supplemented existing technical and environmental codes with labour standards drawn from national laws and international standards, such as ILO conventions. Since 1992, labour codes have mushroomed in branded consumer goods sectors such as food (Diller, 1999). There is no reliable data on the exact number of companies adopting codes. A 1998 study of the International Labour Organization (ILO) documented over 200 company codes relating to worker welfare specifically, with over 20 codes applied to agriculture in developing countries (Mayne, 1999; Blowfield, 2000; Urminsky, undated). In 2001, MEP Richard Howitt stated that he had examples of over 500 codes in his office (Jenkins et al, 2002). In 2005, Neil Kearny, head of the International Textile, Garment and Leather Workers' Federation, cited an estimate of 10,000 individual company codes globally at a European Union conference on responsible sourcing.[16] There are also no accurate figures of the growth of company codes within the food sector. Many of Europe's largest retailers now have codes of labour practice for their suppliers that require compliance with ILO conventions (Henderson, 2002). These include Royal Ahold's Albert Heijn (The Netherlands); Carrefour (France); Kesko (Finland); Migros (Switzerland); and Marks & Spencer, Waitrose, J. Sainsbury and Tesco (UK). These anecdotal estimates and examples give an indication of the growth of company codes of labour practice across all sectors, including food.

Sectoral codes

Around the same time as the development of individual company codes, industry-wide associations became aware that consumer politics could not only damage an individual company, but could threaten the industry as a whole.

As a result, some Northern enterprise associations developed voluntary codes of practice to cover companies within their industry as a proactive measure to avert potential consumer and NGO campaigns (for example, the Ethical Tea Partnership).[17] Independent collaboration between companies is also increasing in relation to monitoring. Four of the UK's leading retailers – Marks & Spencer, Safeway Stores plc, Tesco and Waitrose – together with Northern Foods, Geest, Rank, Hovis, McDougall Plc (RHM) and Uniq formed Supplier Ethical Data Exchange (SEDEX), a web-based system that allows companies to maintain data on the labour standards of their production sites and to make it available to those companies with which they are in a trading relationship (PWC, 2004). A number of producer associations in the South have also developed their own codes as a way of protecting the image and the legitimacy of their industries in Northern markets (see Box 1.2 for examples from African horticulture). In both cases, the industry-wide approach to codes is premised on the notion that companies in an industry or region face similar external pressures and that a coordinated approach will facilitate cooperation between all member companies (Sethi, 2002).

Box 1.2 Sectoral codes in African horticulture

A number of sectoral codes of labour practice have been established through consortia of trade associations and producers in Africa. They moved early to introduce their own standards to promote ethical production. These include the following codes:

* Kenya Flower Council (KFC);
* Fresh Produce Exporters Association of Kenya (FPEAK);
* Zambia Export Growers Association (ZEGA);
* Horticultural Promotion Council (HPC), Zimbabwe.

In both Kenya and Zambia, the horticultural export associations have also participated in the European-Africa-Caribbean-Pacific Liaison Committee (COLEACP) regional harmonized framework, which provides guiding principles for local codes covering product safety and environmental and social responsibility.

Source: adapted from Barrientos, Dolan and Tallontire (2003)

Multi-stakeholder codes

Despite the growth of company and industry codes, their content and process of implementation is far from uniform. Codes differ significantly in terms of the criteria that they contain, who is involved in developing them, the level of communication and capacity building to support them, and how they are monitored. Without independent verification, there is no means of accountability for voluntary company codes, which leaves them open to accusations of 'green washing' by trade unions and NGOs. Since the late 1990s, multi-stakeholder initiatives developed by coalitions of companies, trade unions, NGOs and other civil society stakeholders have evolved in the North and South. These provide the basis for more independent forms of code implementation, monitoring and verification.

Two multi-stakeholder initiatives have been established in the North that are relevant to a number of sectors, including the food industry: the UK Ethical Trading Initiative (ETI) and the US Social Accountability International (SAI).[18] The ETI was established in 1997 as a coalition of companies, trade union organizations and NGOs. It grew from a company membership of 25 in 2000 to 38 companies in 2005. Its membership includes six leading UK supermarkets and nine food companies (ETI, 2005).[19] The ETI has a baseline code that its member companies adopt as the foundation of their own individual company codes of labour practice (see Box 1.3). The ETI is not a code certification body, but a learning organization that facilitates code implementation by its members.

Box 1.3 Summary of the provisions of the Ethical Trading Initiative (ETI) Base Code

- Employment is freely chosen.
- Freedom of association and the right to collective bargaining are respected.
- Working conditions are safe and hygienic.
- Child labour shall not be used.
- Living wages are paid.
- Working hours are not excessive.
- No discrimination is practised.
- Regular employment is provided.
- No harsh or inhumane treatment is allowed.

These provisions are based on internationally agreed labour standards set by the International Labour Organization (ILO) and on other relevant international standards. Employers are also expected to comply with national and other applicable law. The Base Code can be seen in full on ETI's website at www.ethicaltrade.org.

Source: www.ethicaltrade.org

Social Accountability International (SAI) was developed in 1997 by the Council on Economic Priorities (CEP) in consultation with various stakeholders.[20] It promotes SA 8000, which is an independent certifiable code against which suppliers can be audited. The organization has ten retailer 'members', including Dole Foods, who are expected to encourage their suppliers to seek certification (Mosher, 2003). In 2005, SAI issued 710 certifications of SA 8000, of which 7.3 per cent were in food and agriculture, including large multinationals such as Dole and Chiquita.

Other multi-stakeholder initiatives bring together companies, trade unions and NGOs from a single sector, such as the International Cocoa Initiative, which focuses on the elimination of abusive child labour practices in cocoa cultivation and processing (see the Glossary to this volume). Multi-stakeholder initiatives involving companies, trade unions and NGOs have also begun to be established in the South. These include the Wine Industry Ethical Trade Association (WIETA) in South Africa, the Agricultural Ethics Assurance Association of Zimbabwe (AEAAZ) and the Horticultural Ethical Business Initiative (HEBI) in Kenya (see Chapter 8). In addition to voluntary codes, Global Union Federations have begun to develop International Framework Agreements (IFAs) with multinational companies in their sectors. These commit the multinational company to comply with national legislation and ILO conventions in their supply chain, even if they outsource production. They provide a framework for guiding and setting policy and negotiating issues at a local level, where local practice often varies. The International Union of Food, Agricultural, Hotel, Restaurant, Catering, Tobacco and Allied Workers' Associations (IUF)[21] had signed four framework agreements by 2002, including with Danone and Chiquita (Leipziger, 2003). IFAs complement voluntary company codes and provide the basis for adapting industrial relations to the new realities of global sourcing.

Challenges and opportunities for ethical sourcing?

As ethical sourcing increasingly moves into the mainstream of food retailing, new opportunities and challenges are emerging. The opportunities are reflected in the rapid growth of both fair trade and codes of labour practice, with more companies adopting ethical sourcing as part of their mission, and greater consumer awareness of both fair and ethical trade. The challenges are linked to this, as growth brings to the fore tensions between the social goals of fair and ethical trade, and the commercial goals and practices of mainstream food retailing. While the success of fair trade has inspired more mainstream food retailers to adopt ethical sourcing (OPM/IIED, 2000), there is a risk that the principles of ethical sourcing are being watered down through growth. As company codes become more established, contradictions between the social principles that they embody and the commercial goals of corporates for low-cost and fast turnover production are becoming more evident (Acona, 2004; Oxfam, 2004). As ethical sourcing becomes increasingly mainstreamed, it is likely to face new tensions, which we now explore separately for fair and ethical trade.

Challenges for fair trade

The rapid growth of fair trade-labelled products in mainstream commercial retailing raises a number of important issues for the fair trade movement. There is a risk that rapid expansion into fair trade is driven less by ethical objectives than by the commercial imperatives of supermarkets and multinational food companies to differentiate themselves from their competitors and mark out new product areas as part of their commercial growth strategy. This risk is heightened in the case of supermarket fair trade branding of their own-label products and the entry of large multinationals such as Nestlé into the selling of fair trade lines.

However, these risks are complex and can vary by company. As Croft (see Chapter 4) illustrates with regard to the UK Co-operative Group, supermarkets and food companies have different corporate philosophies underpinning their commercial operations and do not necessarily have a uniform commitment to the principles of fair trade. Some, such as the Co-operative Group, have a long corporate tradition based on philanthropy and cooperative values that are similar to those of fair trade. It has, for instance, led the expansion of fair trade among UK supermarkets with its fair trade-retailed products, expanding from a value of UK£100,000 in 1999 to UK£11 million in 2002. The Co-operative Group was also, as Croft describes, the first retailer in the UK to institute ethical values throughout its global supply chains, becoming a founding member of the ETI during the 1990s. Yet, Croft argues that achieving ethical supply relations and improved working conditions requires more than supplying consumers with a fair trade product or ordering a quick factory inspection. Rather, his experience with the Co-op suggests that retailers must be proactive in working with their suppliers to promote ethical supply chain relations. Without such commitment and the active engagement of suppliers, there is little chance that retailers can realize the broader developmental goals envisioned by the fair trade movement. Yet, most supermarkets do not share the Co-operative Group's unique history of social commitment or embrace a business model that privileges the ethical values of social responsibility. Rather, they are in the midst of a highly competitive struggle with other global retailers to achieve commercial success and perceive ethical sales as a way of achieving it (Barrientos and Smith, forthcoming).

There is now concrete evidence that ethical consumption provides a significant marketing opportunity for food retailers and manufacturers. In the UK, the 'ethical market' was valued at UK£19.9 billion (US$35.6 billion) in 2002, and growing numbers of consumers now recognize the Fairtrade Mark and understand what it stands for (Co-op/NEF, 2003). In 2005, the UK research institute Market and Opinion Research International (MORI), found that one in every two adults in the UK now recognize the Fairtrade Mark (up from 25 per cent in 2003 and 39 per cent in 2004) and that the number correctly associating the Fairtrade symbol with its accompanying text 'Guarantees a better deal for third world producers' rose significantly, from 42 per cent in 2004 to 51 per cent in 2005. The survey also revealed that four out of five people who purchase fair trade products considered the integrity of an independently awarded label as important (Fairtrade Foundation, 2004b). The last few years have also witnessed increasing awareness of fair trade in other parts of Europe and the US. In France, for example, public recognition of the fair trade concept increased from 9 to 74 per cent from 2000–2005 (Krier, 2005). The US National Coffee Association reported that recognition of fair trade brands among coffee consumers 18 years and older rose from 7 to 12 per cent from 2003 to 2004, with consumption among this group increasing from 28 to 45 per cent from 2003 to 2004 (Nicholls and Opal, 2005).

Rising consumer awareness is an important part of the success of the fair trade movement in extending its reach and spreading its message. However, it also means that fair trade is extending into new consumer groups who are not necessarily as committed to the fair trade message. They may be purchasing fair trade goods for more hedonistic reasons, such as fashion, or simply because they enjoy the taste or look of the product irrespective of the social philosophy underlying its production. The numbers of these consumers are likely to increase as fair trade becomes mainstreamed in more large supermarkets and retailers, posing new challenges to the long-term viability of its founding ethical principles. As more supermarkets and large food producers expand their own lines of fair trade products, will their commercial values and practices undermine fair trade's ethical principles?[22]

This possibility is generating concern among ATOs. Under the current fair trade model, companies only have to be licensed to carry the Fairtrade logo if they are involved in the production or packaging of a product. While supermarkets purchase coffee, cocoa and other commodities from producers on the FLO register, they do not produce or package food, and are not required to be licensed to carry the logo (Barrientos and Smith, forthcoming). They have therefore made no explicit commitment to a stable purchasing arrangement, a core principle of fair trade. They are able to switch between registered producers at will or could abandon the trading relationship in the event of a market decline. There is also the risk that supermarkets and brand manufacturers will engage in a race to the bottom, 'with buyers scouring the world for the cheapest' fair trade produce (Tallontire and Vorley, 2005, p9). In other cases, mainstream retailers and food manufacturers have opted to have their preferred suppliers inspected and certified by FLO rather than purchase from suppliers who are already on the FLO register (Nicholls and Opal, 2005). Yet, in many cases their existing suppliers are large producers who in no way represent the marginalized small producer whom the fair trade movement seeks to assist. Suppliers who are already integrated within the existing supply chains of major food retailers and manufacturers are by no means the most exploited developing producers. Concerns, for example, have been raised with regard to Nestlé's fair trade-certified coffee brand in the UK, with critics such as the World Development Movement (WDM) claiming that Nestlé's fair trade strategy is actually 'an attempt to cash in on a growing market' rather than 'a fundamental shift in Nestlé's business model' (WDM, 2005).

Likewise, there are now certification schemes, such as US-based Rainforest Alliance and Utz Kapeh (see Box 1.4), that combine some of the principles of fair and ethical trade at a lower cost than full FLO certification.[23] Such schemes have also prompted criticism from within the fair trade movement. Representatives of the Max Havelaar Fairtrade label, for example, have expressed concern that Utz Kapeh represents a form of 'Fair trade-Lite', the response from Utz Kapeh being that they are providing wider access for producers than fair trade (Tallontire and Vorley, 2005). Similar concerns have surrounded Rainforest Alliance, whose certified coffee is used in Kraft, Lavazaa and Proctor and Gamble blends. Such concerns are not entirely baseless. Whereas the Fairtrade Foundation guarantees farmers a minimum price of US$1.21 (UK£0.65) per pound of green coffee beans, Rainforest Alliance, for example, offers no minimum or guaranteed price (Nicholls and Opal, 2005). At the same time, manufacturers such as Kraft and Proctor and Gamble can apply the Rainforest Alliance logo on their packaging if their coffee contains a minimum of 30 per cent certified coffee beans and can thereby capitalize on the growing ethical market (McAllister, 2004). Responding to criticism, Kraft and others have argued that fixed producer prices are antithetical to free market principles and that by introducing the concept of sustainability into existing trade relations Kraft offers small producers a better chance of growing their market in the future. These schemes also widen producers' acess to ethical sourcing beyond niche markets. As the chapters in this book illustrate, the debate between mainstream commercial growth and the integrity of fair trade's founding principles is not easily reconciled and is likely to continue for years to come.

Challenges for ethical trade

The rapid expansion of codes of labour practice since the early 1990s as a means of enhancing workers' rights and employment conditions within the global food system has also posed challenges. These challenges partly relate to the proliferation of codes in recent years, as well as the difficulties of implementing codes in the context of a largely seasonal and often informal agricultural labour force, a significant proportion of which is female (Dolan and Sorby, 2003; Hurst et al, 2005). Producers often face a plethora of codes that cover a range of standards for environmental, food safety and social welfare. As Smith and Dolan's case study (see Chapter 5) shows, suppliers of export produce from Africa have been inundated with codes of labour practice, coming from supermarkets and their agents, multi-stakeholder bodies, and local and international industry associations. Yet, despite the prevalence of codes in the food industry, there are a number of questions about whether they actually reach or benefit the workers they are aimed at. In recent years, awareness has grown that while it is relatively easy to develop a social code, it is more difficult to ensure that it is implemented and monitored in a way that

---

Box 1.4 Preferred supplier schemes: Utz Kapeh

The Utz Kapeh code for 'certified responsible coffee' was developed by a foundation with headquarters in both The Netherlands and Guatemala, with the support of the global retailer Ahold. The code is based on the Euro Retailer Producer Working Group Standard for Good Agricultural Practice (EurepGAP) and is thus a baseline assurance of good agricultural practices in coffee production, but is more detailed and onerous than the normal EurepGAP standards with regard to worker welfare. The foundation aims to 'bring social and environmental performance to the mainstream', and the proposed code for good practice is intended to be 'a ticket to entry to an emerging market for mainstream certified responsible coffee'. Utz Kapeh sets out to support existing brands rather than stake itself in the market as a separate brand. Ahold's Albert Heijn chain of supermarkets in The Netherlands now uses Utz Kapeh certified coffee in all of its own-brand 'Perla' coffees.

Utz Kapeh aligns itself with corporate social responsibility (CSR) and strongly differentiates itself from fair trade, which it views as a niche market. Like other codes of conduct, it does not specify a floor price or a 'living wage'; but the code includes a commitment to long-term commercial relationships between buyers and producers. Utz Kapeh certifies large estates as well as co-operatives of smallholders. The foundation recommends that buyers pay a sustainability differential to producers when market prices are low. These are not fixed and, at present, it is not clear if this is monitored. List prices paid by Utz Kapeh buyers cited by Renard (2004) were above market prices (US$0.70–$0.77 per pound compared with US$0.60 per pound), but significantly below the Fairtrade Labelling Organizations International (FLO) price of US$1.26 per pound.

Yet, an important difference between Utz Kapeh and fair trade is that the former is open to all kinds of producers and producer groups (i.e. large and small, co-operative and estate), and it takes a broader approach to fair trade, aiming to create access to mainstream markets and to increase the competitiveness of participating producers over time.

Source: adapted from Tallontire and Vorley (2005)

---

creates better conditions for workers in global supply chains. Several concerns have been raised regarding the process of code development, the extent to which codes cover all workers in a supply chain, and the capacity of monitoring mechanisms to identify poor employment practices. With respect to the content of codes, a key issue has been the selection of criteria for inclusion in codes and the nature of actors engaged in the process of code formulation. This relates to the fact that most codes have been developed in the North and have involved little (if any) consultation with workers or Southern stakeholders who represent their interests. The content of codes also varies considerably. Most codes call for adherence to national laws and regulations and include core ILO conventions. However, national regulations differ significantly across countries and few independent company codes recognize all of the core ILO conventions. Hence, because there are no universally accepted criteria for inclusion in codes, the content can range from being relatively comprehensive to rather insubstantial.

There are also differences in who codes cover and how far along the supply chain the standard is applied. As Smith and Dolan's (Chapter 5) case study of codes of conduct in the African horticulture industry shows, for codes to deliver substantive changes in working conditions, they need to ensure that the rights of all workers are protected, including those of marginal workers. According to Smith and Dolan, while codes often extend to a company's first-tier suppliers and their permanent employees, they rarely cover all workers in the supply chain, such as small producers, subcontractors and vulnerable workers who may be casual, contract or migrant. The horticulture sector, for example, is characterized by high levels of insecure (informal) employment, the majority of which is occupied by women. Women often have particular concerns to which codes are insensitive: embedded gender discrimination in hiring and promotion of workers; poor conditions linked to the insecurity of women's employment; lack of adequate provision for women workers' reproductive health and child care responsibilities; and poor safety or transport provision for women working overtime (see Chapter 5). In most cases, neither the content of codes nor the depth of their coverage addresses these concerns. Smith and Dolan argue that these sorts of issues can only be uncovered and, ultimately, tackled through gender-sensitive auditing methods, which engage all workers, as well as stakeholders who represent their interests, in the process of code implementation and remediation.

Applying labour codes in supply chains involving vulnerable workers poses important challenges that are also addressed in Prieto-Carrón's study (see Chapter 6) of ethical sourcing in the Central American banana industry, an industry that has long had a reputation for poor labour practices. Like African export horticulture, women working in banana production often face particularly difficult conditions. Over the last decade, multinationals engaged in banana production have come under increasing pressure to improve the labour conditions on their own plantations, as well as on the production units of their subcontractors. Chapter 6 provides an analysis of these efforts through a case study of Chiquita's sourcing operations in Nicaragua and Costa Rica. While Chiquita has a comprehensive policy on corporate social responsibility (CSR) and has taken positive steps to improve worker welfare, Prieto-Carrón's interviews with women workers reveal the challenge of translating corporate policy into concrete and sustainable changes in workplace practice. In particular, her chapter reveals how critical an awareness of gendered social relations is to a company's ability to improve working conditions and its overall performance on corporate responsibility.

An important emerging issue stemming from these studies is the relation between ethical sourcing and the purchasing practices of global buyers, and specifically the extent to which producers are able to genuinely improve labour conditions in the face of the commercial pressures that supermarket buyers exert. Whilst supermarket CSR managers now require compliance with the criteria in codes of labour practice (such as increasing pay and reducing overtime), buyers from the same supermarkets often exert commercial pressures that have the opposite effect (falling prices and shorter lead times). For instance, although producers of fresh produce may receive 'programmes' for predicted purchases in the coming year, firm orders are often not guaranteed until days before the product needs to be on the shelf. This leaves little 'lead time' to fulfill orders to the right specification, or produce being rejected once it reaches distribution centres on spurious grounds. For perishable products this creates enormous problems, especially when they need to be shipped half way around the world. Supermarkets use such 'just-in-time' strategies to squeeze all possible costs out of the chain in order to attract customers through lower prices, while increasing returns to investors (Oxfam, 2004).

Most of these costs are deflected onto suppliers, who are not only expected to absorb the costs of production, but also of promotions and innovations such as new packaging, as well as the costs of regular audits to demonstrate compliance with supermarket standards.[24] While these sorts of practices give supermarkets a market edge in a competitive retail environment, they have also created a constant downward price trend for producers. In the UK, there have been an increasing number of denouncements by campaigning organizations and ex-suppliers about what are deemed unfair purchasing practices by supermarkets and brand marketers, and the effects of these practices on the working conditions of their overseas suppliers (Acona, 2004; Oxfam, 2004; ActionAid, 2005).

Many suppliers accommodate price reductions and cost pressures by increasing the informalization and intensification of work (Barrientos and Kritzinger, 2004; Precision Prospecting, 2005). One example is the UK gangmaster system described by Pollard, which now provides over half of the labour force of the UK's food and farming industries (see Chapter 7). Pollard describes how changes in agricultural production and food marketing have led to a shift away from the use of seasonal localized labour to the year-round use of gangworkers in large-scale production in farms, greenhouses and packhouses. Gangmasters operate as independent third-party employers who provide workers to producers as required, facilitating a highly mobile low-cost labour force. Some gangmasters are reputable labour providers who employ legal migrant workers. But to keep labour costs down, or meet sudden fluctuations in labour demand, a subcontracting system operates where smaller and illicit gangmasters are often used. These gangmasters frequently employ illegal migrant labour, which facilitates an environment of labour abuse. Pollard describes the evolution of different initiatives that have been introduced to address gangmaster abuse. He shows that voluntary initiatives have helped to raise awareness and bring actors together to confront the issue. But the severity of the problem has ultimately required the introduction of legislation to curb the illegal activities practiced by some gangmasters.

Even when voluntary initiatives such as codes of labour practice are firmly in place, the commercial realities of the modern global food system can pose a significant challenge to the realization of decent working conditions. Supermarkets monitor their codes of labour practice through a combination of supplier self-assessment and social auditing. However, supply chains are often complex and can involve different layers of packhouses, producers and subcontractors. Monitoring usually only takes place at the first-tier level, involving processing and/or packaging, and where risk to the supermarket brand name is highest.[25] Yet, it is at the level of production, where the use of casual, seasonal and migrant labour is commonplace, that labour conditions are often poorest. As various chapters in this volume highlight, codes of labour practice are often ineffective for the large temporary, seasonal and migrant workforce involved in agricultural and food production. This has raised questions about the effectiveness of current monitoring procedures for codes of labour practice.

Auret and Barrientos (see Chapter 8) critically examine the most commonly adopted approach to monitoring: a compliance-based approach using snap-shot social auditing and assessments. Such standard auditing is increasingly criticized for providing a 'top-down' approach that fails to pick up or address the genuine needs of workers themselves, whose involvement in an audit is often peripheral (CCC, 2005). Various initiatives have emerged to address the weaknesses of conventional auditing. Social Accountability in Sustainable Agriculture (SASA), for example, was a collaborative project developed by FLO, SAI, the Sustainable Agriculture Network (SAN) and the International Federation of Organic Agriculture Movements (IFOAM), four main social and environmental verification systems in sustainable agriculture, to improve social auditing processes in agriculture and to foster shared learning between the participating initiatives.[26] Auret and Barrientos (Chapter 8) explore another alternative to standard auditing practice: participatory social auditing (PSA). PSA provides a process approach to social auditing, in which the inclusion of workers themselves is central, and draws on participatory tools such as focus group discussions, role play and ranking exercises to facilitate worker participation. However, participation needs to operate at different levels to ensure that the voices of insecure workers are included. Trade unions, NGOs and various stakeholders are now calling for codes to be verified and monitored by independent monitoring bodies, and for greater participation of Southern stakeholders. Auret and Barrientos examine local multi-stakeholder initiatives that have involved trade unions and NGOs in monitoring and verification that are engaged in the issues facing workers.

The chapters in this book highlight the benefits of ethical trade in raising the issues and bringing different actors together to address workers' rights in the context of the global food system. But they also highlight the limitations of voluntary initiatives given the complexities and pressures of global sourcing. As the number of code initiatives continues to expand, the tension between ethical sourcing and the commercial dynamic of the global food system is likely to intensify. This raises questions as to whether a consumer-focused food system based on the constant provision of cheap food is compatible with voluntary initiatives such as ethical trade. NGOs as well as trade unions are now questioning whether voluntary initiatives are sufficient to address workers' rights. International framework agreements with multinational companies adopted by the IUF and Global Union Federations in other sectors provide a common framework for collective bargaining across the range of suppliers and countries from which they source. Some NGOs argue that it is only regulation that will curb the adverse effects of multinational sourcing on workers and small producers, and achieve more equitable trade (ActionAid, 2005). How the relationship between voluntary and regulatory approaches plays out poses profound challenges for practitioners, policy-makers, and consumers of ethical trade.

Addressing the challenges and realizing the opportunities

As fair and ethical trade expand, there are increasing issues of overlap between them (Smith and Barrientos, 2005). The growth of fair trade has led to an expansion of its product range, resulting in pressure to source from a larger number of producers. While fair trade's founding principles were oriented towards small and marginalized producers and producer groups, there is now a move towards purchasing from larger commercial farms or 'plantations'. This was spearheaded by FLO's standard for banana certification in 1997, which included coverage of minimum labour standards for workers. The plantation standard has been extended to other products, including fruit and wine. However, the extension of fair trade to 'plantations' and large commercial farms with a significant wage labour force has also created an overlap with the criteria of ethical trade.[27]

There has been resistance from some within the fair trade movement to this trend on the grounds that larger producers will marginalize small-scale and family farms, whose protection is a foundation principle of fair trade. This has been a particular issue in coffee, as coffee plantations are not eligible for fair trade certification in order to protect small-scale producers. South Africa has been able to reconcile these tensions in the deciduous wine and fruit sector (see Box 1.5) through the development of country-specific criteria that certifies larger commercial farms as fair trade. But the degree to which this model can be rolled out more broadly remains an open, yet critical, question to the future of fair trade. Whether fair trade can accommodate larger commercial farms without compromising small producers depends partly upon its future trajectory and growth patterns.[28]

The debate over 'plantations' underpins one of the key issues facing fair trade. It faces tremendous opportunities in terms of growth, which could take it out of a small niche market, and influence the rise of more equitable trade within mainstream sourcing. But to meet consumer demands for a growing range and quantity of fair trade products, fair trade is likely to expand further

Box 1.5 Fair trade in South African deciduous fruit and wine

In South Africa there has been a growing trend towards fair trade in the deciduous fruit and wine sector, a sector that has expanded significantly following the transition to democracy after 1994. Land tenure in commercial agriculture is characterized by medium-sized commercial farming, with a predominantly white ownership. This system resulted from the apartheid policy of segregation, which forced smaller non-white producers off the land. Coloured producers became wage labourers living on white-owned farms and black Africans were largely forced into the 'homelands' forming a pool of seasonal migrant labour.

The transition to democracy facilitated the opening-up of the fruit and wine sectors to export markets, and producers selling to supermarkets were expected to meet minimum labour standards specified by South African law. As part of initiatives to encourage black economic empowerment, some producers adopted equity share schemes for their workers and local communities. A group of these producers applied for fair trade certification to facilitate market access for their Thandi brand. In the context of a difficult commercial environment in 2002, many other South African producers followed suit and sought to become fair trade-labelled based purely on their compliance with South African law. This led stakeholders in South Africa to argue that the Fairtrade Labelling Organizations International (FLO) standards needed to be adapted to the South Africa context, making land reform and black economic empowerment an explicit aim. Following extensive discussion, FLO agreed to introduce conditions ensuring that only producers with a strategy for black empowerment could become fair trade labelled. Hence, fair trade labelling has been adapted to the specific context of 'plantation' agriculture in South Africa.

Source: adapted from Barrientos and Smith (forthcoming)

into larger commercial farms that employ wage labour. As it does so, to what extent can fair trade be differentiated from other forms of ethical sourcing such as Ethical Trade? Do FLO and national fair trade initiatives have the ability to maintain the principles of ethical sourcing across a wider tranche of commercial trade? Or will the commercialization of fair trade drive a wedge through the fair trade movement? These are not easily resolvable questions. They arise as a consequence of the success of fair trade in expanding into mainstream food production and retailing. The tensions between the social aims of fair trade, and the commercial realities of a competitive global food system are likely to continue to characterize the movement over the foreseeable future. Successfully managing these tensions will remain a significant challenge.

Ethical trade faces similar challenges, coming from a different direction. The proliferation of codes and the ability of NGOs to influence large corporations are signs of the success of ethical trade. But sections of the movement are now addressing the tension between the social principles of codes and the commercial realities of the global food system. Some NGOs and trade unions are arguing for more ethical sourcing practices to be implemented by multinational companies in their own purchasing activities (such as better communication flows with suppliers, more stable commercial relationships, paying higher prices and giving longer lead times). Does this bring ethical trade into the realm of fair trade? Is it possible to have ethical trade without a wider set of policies that improve equity in mainstream trading relations?

Addressing ethical sourcing in mainstream trade

The success of both fair trade and ethical trade over the past decade is striking. They have both witnessed rapid growth. An increasing number of consumers are now aware of the social issues associated with the production of the goods they purchase. Fair and ethical trade are now in a position to influence the conditions of a growing number of producers and workers within the global food system. But growth has also brought with it significant challenges, particularly as this growth confronts the commercial realities of global trade. Despite increasing interest in ethical sourcing, reversing the structural inequalities of global trade remains a far-off vision. The realities for many producers and workers in a global trading system dominated by large food multinationals and retailers are falling prices and a more insecure commercial environment (Robbins, 2003).

One example is the coffee sector, which has experienced a major crisis over the past decade. The price of coffee fell by almost 50 per cent between 1999 and 2002, resulting in a 30-year low. While prices paid to coffee producers have fallen, the retail price has remained stable, with multinationals at the centre of the chain benefiting from the difference. Campher (see Chapter 9) examines the effect of the crisis on developing country coffee farmers, mostly poor smallholders, who often sell their coffee beans at a heavy loss, while branded coffee realizes a hefty profit. He describes the problems currently facing coffee producers in developing countries and how the power of large multinational coffee processors and distributors has aggravated these problems. He argues that the scale of the solution needs to be commensurate with the scale of the crisis. Initiatives by large corporations such as Starbucks and the entry of Nestlé into the fair trade market demonstrate some movement; but this remains limited. Chapter 9 describes Oxfam's campaign to develop a Coffee Rescue Plan, which brought together all the major players in the coffee trade to support coffee producers and workers. Yet, Campher argues that despite the success of the Coffee Rescue Plan, the issues extend well beyond coffee to the underlying structure of trading relations and the global challenge to make trade fair.

The changing dynamics of the global food system, however, are also reflected in the wider effects of supermarket retailing and centralized procurement on producers and workers. Fox and Vorley (see Chapter 10) analyse the implications of the growing concentration in the processing and retail sectors of regional agri-food systems for rural livelihoods and communities, and question whether small farmers and food industry enterprises can survive at all in the globalized agri-food system. Their chapter specifically addresses how changes in retail, driven by global, regional and national factors, are affecting the ways in which agri-food chains are organized for commodities that are important to small- and medium-scale farmers and other rural economic agents. In particular, they examine the role of primary producers and their economic organizations in negotiating market access and improving terms of trade in the specific agricultural supply chains of large supermarkets. Fox and Vorley stress the need for producers, policy-makers and civil society to anticipate and respond to the profound transformations occurring in the agri-food markets in which they engage, and raise the need for the regulation of food multinationals and supermarkets. Ultimately, their chapter raises the question of whether fair trade can have a significant effect if there is not a substantial change in the global trading environment itself.

Conclusions

The chapters in this volume bring together a wide range of experiences of ethical sourcing in the global food system. They highlight both the success of NGO and trade union pressure for fair and ethical trade, but also the challenges that these initiatives face. At present, both ethical and fair trade remain fairly niche activities, limited to a small number of small-scale producers or workers at the upper tiers of supermarket supply chains. Both are experiencing challenges, often as a result of their success, which will have to be overcome if their reach is to continue expanding. But the momentum that has been generated by these initiatives is, nevertheless, significant. Only a few visionaries would have thought a decade ago that such initiatives could become established at all in the mainstream of food production and retailing.

If ethical sourcing initiatives are viewed in isolation, they can appear relatively peripheral to the wider issue of poverty reduction. But their contribution has important implications when they are viewed as part of a wider movement aimed at addressing inequities in global trade. Fair and ethical trade are based on loose coalitions of diverse civil society organizations. Many of these actors are also part of the Trade Justice Movement and similar civil society campaigns aimed at addressing the structural constraints to poverty reduction. An analysis of these wider civil society campaigns is beyond the scope of this book, which has focused on voluntary approaches to ethical sourcing. But this movement is playing an important role in challenging existing inequitable trade rules that inhibit developing countries in their trading activities and provide unfair advantages to developed countries and multinational corporations. Despite the challenges facing fair and ethical trade, they highlight the advances that can be (and have been) made through collective action and innovative initiatives. Both approaches have raised consumer awareness, and have shown that commercial success and social justice are not necessarily irreconcilable. This book aims to contribute to a better understanding of ethical sourcing, where we are now, and the challenges that face the movement ahead.

Notes

1   BSE is defined as bovine spongiform encephalopathy, CJD as Creutzfeldt–Jakob disease.
2   See, for example, Bernstein et al (1990); Thrupp (1995); Shiva (2000); Robbins (2003); Lang and Heasman (2004); Lawrence (2004); and Vorley (2004).
3   In addition to the previous references, see, for example, Christian Aid (1997); Brandt (1999); Oxfam (2004); Smith et al (2004); and ActionAid (2005).
4   The environmental and product dimensions, such as organic production, are only touched upon where they relate to or inform an understanding of the social dimensions.
5   FINE is a forum that brings together the following fair trade organizations: Fairtrade Labelling Organizations International (FLO), International Fair Trade Association (IFAT, formerly the International Federation of Alternative Trade), Network of European World Shops (NEWS) and European Fair Trade Association (EFTA). See the Glossary for more detail.
6   The full definition of fair trade can be accessed at www.bafts.org.uk/fair-trade/ fine.htm.
7   The core ILO conventions are: No 87 Freedom of Association and Protection of the Right to Organize Convention; No 98 Right to Organize and Collective Bargaining Convention; No 29 Forced Labour Convention; No 105 Abolition of Forced Labour Convention; No 111 Discrimination (Employment and Occupation) Convention; No 100 Equal Remuneration Convention; No 138 Minimum Age Convention; and No 182 Worst Forms of Child Labour.
8   After a major review of its fair trade sourcing strategy during 1997–1998, Oxfam switched from direct sourcing of fair trade products to retailing products imported by other fair trade organizations for efficiency reasons (Nicholls and Opal, 2005).
9   See the Glossary for a more detailed list of key fair trade organizations and their website addresses. Most of these organizations are part of broader networks/ umbrella organizations, such as IFAT, EFTA and NEWS, which were established by ATOs to facilitate cooperation and information flows between organizations involved in fair trade (see the Glossary) (OPM/IIED, 2000).
10   See the Café Direct website: www.cafedirect.co.uk/fairtrade/fairtrade.php.
11   Fairtrade labelling has three core components: social and environmental standards; a fair price; and market access for disadvantaged producers.
12   Despite rapid growth, at around UK£130 million, fair trade sales account for only 0.17 per cent of the UK£76 billion spent on food and drink in the UK in 2003, or 0.09 per cent of the UK£144 billion total consumer expenditure on food, drink and catering services (Tallontire and Vorley, 2005).
13   Figures released by the UK Fairtrade Foundation in 2003, cited in *Freshinfo News Bulletin*, 7 July 2003, www.freshinfo.com.

14   Within the retail sector, particularly in the UK, own-brand products are now central to marketing strategies, accounting for nearly half of total grocery sales (Burch and Goss, 1999).

15   The McDonald's Corporation began selling fair trade-certified coffee exclusively in 658 of its restaurants across New England and in Albany, New York, in November, 2005.

16   European Commission Conference on Responsible Sourcing: Improving Global Supply Chains Management, held in Brussels on 18 November 2005.

17   Formerly the Tea Sourcing Partnership.

18   There are a number of other Northern, multi-stakeholder initiatives in non-food sectors (e.g. the Fair Labour Association and the Clean Clothes Campaign for garments), but which are not covered here.

19   ETI members relating to the food industry include the International Union of Food Workers; supermarket members ASDA, the Co-op, J. Sainsbury, Marks & Spencer, Safeway, Somerfield and Tesco; and food companies Chiquita, Ethical Tea Partnership, Flamingo Holdings, Fyffes, Premier Foods, Union Coffee Roasters, RGB Coffee Ltd and World Flowers (ETI, 2005).

20   CEP formed a company, the Council on Economic Priorities Accreditation Agency (CEPAA), to promote and validate certification according to SA 8000. This organization has been renamed as Social Accountability International.

21   International Union of Food, Agricultural, Hotel, Restaurant, Catering, Tobacco and Allied Workers' Associations.

22   An additional critique of fair trade has been advocated from a neo-liberal perspective by the Adam Smith Institute. Lindsey (2004) argues that the woes of coffee producers are aggravated by fair trade encouraging overproduction by small holders (Lindsey, 2004). For a discussion of the economics of fair trade, see Nicholls and Opal (2005).

23   Rainforest Alliance, for example, does not charge a licensing fee to use the alliance's logo in contrast to the 2 per cent levied by the Fairtrade Foundation.

24   For a critique of supermarket practices, see, for example, Lawrence (2004); Vorley (2004); Bevan (2005); and Blythman (2004).

25   Monitoring of supermarket codes of labour practice is more common in Europe, particularly the UK, but US supermarkets rarely apply social codes in the supply base.

26   See www.isealalliance.org/sasa/documents/SASA_SummaryUpdate_0305.pdf.

27   See Nicholls and Opal (2005) for a discussion of Afrofair and the development of Fairtrade bananas.

28   Beyond the North, fair trade has recently been expanding its reach in parts of the South, especially Latin America and, to a lesser extent, Asia and Africa. This is a relatively new advance, but provides potential fair trade market opportunities for local producers within their own countries, as well as through exports. There has also been pressure from Northern producers to be given access to Fairtrade retailing; but this has been resisted by the Fairtrade movement, which views it as watering down its principle of providing access for more marginalized Southern producers. Farmers' markets and organic box schemes are the nearest that Northern producers have been able to advance along these lines, although the Organic Soil Association introduced an ethical trade label in 2005 (Tallontire and Vorley, 2005).

References

Acona (2004) *Buying Your Way into Trouble? The Challenge of Responsible Supply Chain Management*, Insight Investment Management Ltd, London

ActionAid (2005) *Power Hungry: Six Reasons to Regulate Global Food Corporations*, ActionAid, London

ANPED (Northern Alliance for Sustainability) (1999) *Report from the NGO Conference From Consumer Society to Sustainable Society: Towards Sustainable Production and Consumption*, The Northern Alliance for Sustainability, Soesterberg, The Netherlands, www.anped.org/media.php?id=20, accessed 11 March 2005

Barratt Brown, M. (1993) *Fair Trade, Reform and Realities in the International Trading System*, Zed Press, London

Barrientos, S., Dolan, C. and Tallontire, A. (2003) 'A gendered value chain approach to codes of conduct in African horticulture', *World Development*, vol 31, no 9, pp1511–1526

Barrientos, S. and Kritzinger, A. (2003) 'The poverty of work and social cohesion in global exports: The case of South African fruit', in Chidester, D., Dexter, P. and Wilmot, J. (eds) *What Holds Us Together: Social Cohesion in South Africa*, HSRC Press, Cape Town

Barrientos, S. and Kritzinger, A. (2004) 'Squaring the circle – Global production and the Informalisation of work in South African fruit exports', *Journal of International Development*, vol 16, pp81–92

Barrientos, S. and Smith, S. (forthcoming) 'Mainstreaming fair trade in global value chains: Own brand sourcing of fruit and cocoa in UK supermarkets', in Murray, D. and Raynolds, L. (eds) submitted for publication

Bernstein, H., Crow, B. Mackintosh, M. and Martin, C. (1990). *The Food Question, Profits Versus People?*, Earthscan, London

Bevan, J. (2005). *Trolley Wars: The Battle of the Supermarkets*, Profile Books, London

Blowfield, M. (1999) 'Ethical trade: A review of developments and issues', *Third World Quarterly*, vol 20, no 4, pp753–770

Blowfield, M. (2000) 'Ethical sourcing: A contribution to sustainability or a diversion?', *Sustainable Development*, vol 8, pp191–200

Blythman, J. (2004) *Shopped: The Shocking Power of British Supermarkets*, Fourth Estate, London

Brandt, D. (1999) *Women Working the Nafta Food Chain: Women, Food and Globalization*, Second Story Press, Toronto

Burch, D. and Goss, J. (1999) 'Notes from global sourcing and retail chains: Shifting relationships of production in Australian agri-foods', *Rural Sociology*, vol 64, pp334–350

CCC (Clean Clothes Campaign) (2005) *Looking for a Quick Fix: How Weak Social Auditing is Keeping Workers in Sweatshops*, CCC, Amsterdam

Christian Aid (1996) *The Global Supermarket, Britain's Biggest Shops and Food from the Third World*, Christian Aid, London

Christian Aid (1997) *Change at the Check-out: Supermarkets and Ethical Business*, Christian Aid, London

Conroy, M. (2001) *Can Advocacy-Led Certification Systems Transform Global Corporate Practices? Evidence and Some Theory*, www.umass.edu/peri/pdfs/WP21.pdf

Co-op/NEF (Co-operative Bank and New Economics Foundation) (2003) *The Ethical Consumerism Report 2003*, Co-op/NEF, www.neweconomics.org/gen/uploads/tkuggj ui2ngnmpvjyyi5hqrw08012004145902.pdf

Diller, J. (1999) 'A social conscience in the global marketplace? Labour dimension of codes of conduct, social labelling and investor initiatives', *International Labour Review*, vol 138, no 2, pp99–129

Dolan, C. and Humphrey, J. (2004) 'Changing governance patterns in the trade in fresh vegetables between Africa and the United Kingdom', *Environment and Planning*, vol 36, no 3, pp491–509

Dolan, C. and Sorby, K. (2003) *Gender and Employment in High-Value Agriculture Industries*, World Bank, Washington, DC

EFTA (nd) 'Fair trade definition', www.eftafairtrade.org/definition.asp, accessed 23 February 2006

Enloe, C. (1990) *Bananas, Beaches, and Bases: Making Feminist Sense of International Politics*, UC Press, Berkeley and Los Angeles

ETI (Ethical Trading Initiative) (2003) *ETI Workbook: Step-by-Step to Ethical Trade*, ETI, London

ETI (2004) *Inspecting Labour Practice in the Wine Industry of the Western Cape, South Africa, 1998–2001*, Ethical Trading Initiative, London

ETI (2005) *ETI Annual Report 2004/2005 – Ethical Trading Initiative: Driving Change*, ETI, London

Fairtrade Foundation (2004a) www.fairtrade.org.uk/about_sales.html

Fairtrade Foundation (2004b) www.fairtrade.org.uk/pr150504.htm

Fairtrade Foundation (2005) 'Fairtrade fortnight 2005 – Volunteer briefing sheet', www.fairtrade.org.uk/downloads/pdf/fortnight_supermarket_briefing_sheet.pdf

FAO (United Nations Food and Agriculture Organization) (2004) 'Improving the quality and safety of fresh fruits and vegetables: A practical approach', in Piñeiro, M., Berania, L. and Ríos, D. (eds) *Food Quality and Standards Service, Food and Nutrition Division*, FAO, Rome

FLO (2005) *Delivering Opportunities: Annual Report 2004/5*, FLO, Brussels

FLO (2006) www.fairtrade.net/sites/impact/facts.htm, accessed 23 February 2006

Grimes, K. M. and Milgram, B. L. (eds) (2000) *Artisans and Cooperatives: Developing Alternative Trade for the Global Economy*, University of Arizona Press, Tucson

Henderson Global Investors (2002) *Socially Responsible Investment: Engagement and Research Review Summer 2002*, Henderson Global Investors, www.ethicalinvestors. com/news/henderson0208.pdf

Hughes, A. (2004) 'Accounting for ethical trade: Global commodity networks, virtualism and the audit economy', in Hughes, A. and Reimer, S. (eds) *Geographies of Commodities*, Routledge, London, pp215–232

Hurst, P. with Termine, P. and Karl, M. (2005) *Agricultural Workers and their Contribution to Sustainable Agriculture and Rural Development*, FAO, ILO and IUF, Rome, www. agribusinessaccountability.org/pdfs//335_Agricultural-Workers-&-Sustainable-Agriculture.pdf

Jenkins, R., Pearson, R. and Seyfang, G. (eds) (2002) *Corporate Responsibility and Labour Rights: Codes of Conduct in the Global Economy*, Earthscan, London

Johnston, J. (2002) 'Consuming social justice: Fair trade shopping and alternative development', in Goodman, J. (ed) *Protest and Globalization*, Pluto Press, Annandale, pp38–56

Kocken, M. (2003) *Fifty Years of Fair Trade: A Brief History of the Fair Trade Movement*, www.gepa3.de/download/gepa_Fair_Trade_history__en.pdf

Krier, J. M. (2005) *Fair Trade in Europe 2005: Facts and Figures in 25 European Countries*, Fair Trade Advocacy Office, Brussels

Ladbury, S. and Gibbons, S. (2000) *Core Labour Standards: Key Issues and a Proposal for a Strategy*, Department for International Development, London

Lang, T. and Heasman, M. (2004) *Food Wars: The Global Battle for Mouths, Minds and Markets*, Earthscan, London

Lawrence, F. (2004) *Not on the Label*, Penguin, Harmondsworth, UK

Leipziger, D. (2003) *The Corporate Responsibility Code Book*, Greenleaf, Sheffield, UK

Levi, M. and Linton, A. (2003) 'Fair trade: A cup at a time?', *Politics and Society*, vol 31, no 3, pp407–432

Lindsey, B. (2004) *Grounds for Complaint? Fair Trade and the Coffee Crisis*, Adam Smith Institute, London

Mayne, R. (1999) 'Taking the sweat out of sweatshops', www.unesco.org/courier/1999_11/uk/ethique/txt1.htm

McAllister, S. (2004) 'Who is the fairest of them all?', *The Guardian*, Wednesday, 24 November, www.guardian.co.uk/ethicalbusiness/story

Moore, G., Gibbon, J. and Slack, R. (2004) 'Social enterprise and fair trade: The challenges of the mainstream', *Mimeo*, Newcastle Business School, Newcastle, UK

Mosher, S. (2003) 'Corporate codes of conduct: A summary', *China Rights Forum*, vol 1, pp43–48

Nicholls, A. and Opal, C. (2005) *Fair Trade: Market-Driven Ethical Consumption*, Sage, London

OPM/IIED (Oxford Policy Management/ International Institute for Environment and Development) (2000) *Fair Trade: Overview, Impact, Challenges – Study to Inform DFID's Support to Fair Trade*, OPM/IIED, Oxford and London

Oxfam (2004) *Trading Away Our Rights: Women Working in Global Supply Chains*, Oxfam International, Oxford

Ponte, S. (2002) *Standards, Trade and Equity: Lessons from the Specialty Coffee Industry*, Centre for Development Research, CDR Working Paper, 13 February, Copenhagen

ProspectingPrecision (2005) *Temporary Workers in UK Agriculture and Horticulture*, ProspectingPrecision, Framlington, Suffolk

PWC (Price Waterhouse Cooper) (2004) *Sourcing Overseas for the Retail Sector: CSR and the Ethical Supply Chain*, PWC, www.pwcglobal.com/uk/eng/about/svcs/PwC-SourcingOverseasPt2.pdf

Ransom, D. (2001) *No Nonsense Guide to Fair Trade*, Verso, London

Raynolds, L. (2000) 'Re-embedding global agriculture: The international organic and fairtrade movements', *Journal of Agriculture and Human Values*, vol 17, pp297–309

Raynolds, L. (2002) 'Poverty alleviation through participation in fair trade coffee networks: Existing research and critical issues', Background paper prepared for project funded by the Community and Resource Development Program, Ford Foundation, www.colostate.edu/Depts/Sociology/FairTradeResearchGroup/doc/rayback.pdf

Reardon, T. and Berdegué, J. (2002) 'The rapid rise of supermarkets in Latin America: Challenges and opportunities for development', *Development Policy Review*, vol 20, no 4, pp371–388

Renard, M.-C. (2004) 'Between equity and the market: Fair trade', paper presented at the XI World Congress of Rural Sociology, Trondheim, 25–30 July

Robbins, P. (2003) *Stolen Fruit: The Tropical Commodities Disaster*, Zed Press, London

Sengenberger, W. (2002) *Globalisation and Social Progress: The Role and Impact of International Labour Standards*, Fredrick Ebert Stiftung, Bonn

Sethi, P. S. (2002) 'Corporate codes of conduct and the success of globalization', *Ethics and International Affairs*, vol 89, no 18, pp89–106

Shiva, V. (2000) *Stolen Harvest: The Highjacking of the Global Food Supply*, Zed Press, London

Smith, S., Auret, D., Barrientos, S., Dolan, C., Kleinbooi, K., Njobvu, C., Opondo, M. and Tallontire, A. (2004) 'Ethical trade in African horticulture: Gender, rights and participation', *IDS Working Paper 223*, Institute of Development Studies, Brighton

Smith, S. and Barrientos, S. (2005) 'Fair trade and ethical trade: Are there moves towards convergence?', *Sustainable Development*, vol 13, pp190–198

Tallontire, A. and Vorely, B. (2005) 'Achieving fairness in trading between supermarkets and their agrifood supply chains', UK Food Group Briefing, www.agribusinesscenter. org/docs/Marketplace_6.pdf

Taylor Nelson Sofres (2005) www.tiscali.co.uk/news/newswire.php/news/reuters/2005/ 01/13/business/tescograbsmoreofmarketinsupermarketwar.html&templ

Thrupp, L. (1995) *Bittersweet Harvests for Global Supermarkets, Challenges in Latin America's Agricultural Export Boom*, World Resources Institute, Washington, DC

Traidcraft (2003) 'European fair trade market overview', Traidcraft, www.traidcraft. co.uk/temp/rad418C0.pdf, accessed 11 March 2005

Transfair (2005a) *TransFair USA Joins Oxfam in Welcoming McDonald's Rollout of Fair Trade Certified (TM) Coffee*, Press Release, Transfair, www.transfairusa.org/content/ about/pr_051031.php

Transfair (2005b) *2005 Fair Trade Coffee Facts and Figures*, Transfair, www.transfairusa. org/content/Downloads/2005Q2FactsandFigures.pdf

Urminsky, M. (ed) (undated) 'Self regulation in the workplace: Codes of conduct, social labelling and socially responsible investment', *MCC Working Paper*, no 1, ILO, Geneva

Vorley, B. (2004) *Food Inc.: Corporate Concentration from Farm to Consumer*, London, UK Food Group

WDM (World Development Movement) (2005) 'Statement by WDM on Nestlé Fairtrade Partner's Blend Coffee', WDM, www.wdm.org.uk/news/presrel/current/ nestle.htm

Wells, P. and Jetter, M. (1991) *The Global Consumer: Best Buys to Help the Third World*, Victor Gollancz, London

Young, G. (2003) *Fair Trade's Influential Past and the Challenges of Its Future*, Report prepared for the Fair Trade Conference: An Asset for Development, held by the King Baudouin Foundation, Brussels

2

The Development of Alternative and Fair Trade: Moving into the Mainstream

Anne Tallontire

What is fair trade?

The profile of fair trade is becoming increasingly prominent. It is an approach to trade that has a strong development rationale, based on introducing previously excluded producers to potentially lucrative markets, building up the capacity of producers to trade effectively in the market or offering a good price to small producers. Fundamentally, fair trade aims to benefit primary producers and attempts to sell their produce to a niche market of consumers who are willing to buy goods that are identified as 'fair trade' and for the benefit of the producer, often at a premium price.

In Northern Europe, the range of fair trade products available continues to expand, and more conventional companies offer a product with a fair trade label. In Europe alone the annual retail sales of fair trade products increased from 260 million Euro in 2001 to 660 million Euro in 2005 (Krier, 2005).[1] While this is less than 0.01 per cent of global trade, fair trade is growing. As recently as the early 1990s, fair trade was a relatively unknown movement operating at the margins of development and business. However, fair trade labelling was then beginning to revolutionize how fair trade operated and widen its horizons to more mainstream outlets and partners. Now, the sales of fair trade-labelled products are growing rapidly,[2] and fair trade claims an influence far beyond its market share. But what does fair trade mean, and what are the origins of the organizations and products that claim to be offering a better deal to developing country producers?

Origins and approaches to fair trade

Two different types of fair trade organization have emerged: namely, *alternative trade organizations* (ATO), followed by *fair trade labelling organizations*. The first ATOs began to operate during the 1950s and 1960s, purchasing goods from disempowered producers with a view to promoting their development as part of goodwill selling and, later, solidarity trade. Prominent examples of ATOs include Traidcraft in the UK and Fair Trade Organization (FTO, formerly SOS Wereldhandel) in The Netherlands. Of the fair trade labelling organizations, the first and probably best known is Max Havelaar in The Netherlands and the Fairtrade Foundation in the UK. Max Havelaar and the Fairtrade Foundation are just two of 20 national initiatives making up Fairtrade Labelling Organizations International (FLO), which was formed in 1997 to coordinate the growing number of labelling organizations.[3]

The brand approach to fair trade

Today's ATOs focus on offering a trading partnership with producers. Their unique selling point or brand is their relationship with producers. Many ATOs have evolved from charities that tried to assist producers by buying their goods; but some have been established explicitly as fair traders, including the new wave of fair trade companies such as Cafédirect, which have begun to bring fair trade to a more mainstream audience. While using trade as a tool for development has remained a constant focus, ATOs have changed dramatically from their origins in 'helping by selling' to more professional and commercially focused operations. We can identify the following broad periods of ATO operations:

*   *Goodwill selling, mid 1950s to early 1970s.* This was a relatively innocent period that typically began with non-governmental organizations (NGOs) selling goods produced by people with whom they were working on development or relief projects. These trading links were often initially on an *ad hoc* basis, with the gradual development of more solid relationships and were usually focused on craft products. Pioneers in Europe included SOS Wereldhandel and Oxfam Trading; the latter began by selling pincushions.
*   *Solidarity trade, 1970s to late 1980s.* While many trading relationships developed in the period of goodwill selling remained, during the 1970s ATOs began to look for new sets of producers. These were typically groups of producers organized collectively or based in countries with governments that explicitly challenged the prevailing economic order (e.g. Nicaragua under the Sandinistas) or supported opponents of apartheid (e.g. Tanzania and other African front line states). The messages to consumers were frequently politically motivated; their purchase was seen as an expression of solidarity with the producer or producing country, such as Nicaragua. While solidarity trading did reach a committed band of

alternative consumers, it had some internal limitations (notably conflicts between campaigning and marketing), and as the international political climate changed, the solidarity message became less tenable. Coffee from Nicaragua or Tanzania is a product typical of this period.

- *Mutually beneficial trade, the 1990s.* The producer focus of earlier periods was associated with the neglect of the consumer. As profits dropped and some ATOs faced bankruptcy, many began to look towards consumer needs and to balance these with the needs of producers. Consumer marketing, product development and product quality all became important concerns of ATOs, marking increased commercial awareness. Despite some internal discomfort with marketing, it came to be seen as a useful tool for the development of the ATO business and the benefit of producers. The key message to consumers was that trade should be mutually beneficial, and ATOs were keen to stress the mutuality of trading relationships to their producers. The volume of food stuff sold on a fair trade basis began to rival crafts.

- *Developing trading partnerships.* The emerging trend from the mid 1990s is a development of the concept of mutual benefits for the producer and consumer into a more clearly defined partnership. ATOs explicitly use the language of partnership both in terms of their relationship with producers and with consumers. Emphasis on partnership has evolved over the years from a more service-oriented relationship; in the past, the ATO provided 'producer services', something for the benefit of the producer. This implied asymmetry in the relationship, with benefits flowing from the ATO to the producer. It is now increasingly recognized that benefits flow from the producer to the ATO; producer testimony lends authenticity to the claims of the ATO and is an important tool in marketing. The concept of mutual benefits in trade evolved and became a broader partnership approach. Food products became increasingly significant relative to craft products.

- *Marketing a fair trade brand.* Espousing the partnership principles that emerged during the early 1990s, a new breed of ATO became more significant during the late 1990s. ATOs such as Cafédirect and the Day Chocolate Company were consciously established to operate in mainstream markets and to use fair trade principles in more conventional company settings, usually specialized in food marketing. By the turn of the 21st century, these companies were at the cutting edge of fair trade mainstreaming.

Most ATOs in the North would claim to be engaged in more equal partnerships with producers in the South. Interestingly, there are now ATOs based in the South, often organizations marketing for groups of primary producers. The International Fair Trade Association (IFAT)[4] is the mouthpiece for ATOs globally and has members in the North and South.

Amongst consumer ATOs in the North, there are interesting differences in approach between countries. In the UK and The Netherlands, where fair trade has a high public profile, ATOs are becoming more oriented towards mainstream markets. For example, in the UK, Traidcraft has worked with supermarkets to enable them to source fair trade own-brand products and has a unit dedicated to helping producers access mainstream markets. By contrast, until recently, fair trade in the US has tended to be more 'purist', confined to specialist fair trade shops catering for an alternative market. To a certain extent, the differing strategies reflect different philosophies of fair trade, with the US ATOs highlighting the symbolic role of fair trade as an alternative to the conventional market.[5] However, the relative concentration of the UK and other European food retail markets has meant that once access to certain retailers has been won, fair trade has gained a large audience for its message and wares. A concern to widen the opportunities in the market for fair trade producers was one of the factors that led to the birth of fair trade labelling initially in The Netherlands, as discussed in the following section.

Most Northern ATOs sell through their own retail channels, including shops, catalogues or the internet. Some, however, specialize in import, selling to other ATOs, including 'world shops', independent shops dedicated to selling fair trade goods, many of which are linked to the Network of European World Shops (NEWS).

Like many charitable organizations such as the Save the Children Fund (SCF) or Médecins Sans Frontières (MSF), ATOs largely rely on their brand reputation to communicate their values – that is to say, you know and trust SCF to work with communities to alleviate poverty through development programmes, and similarly you know and trust a fair trade brand. ATO brands have a high marketing value; but some people in the fair trade movement are pointing out that consumer and producer trust in a fair trade brand does not make them accountable. In order to improve its accountability to stakeholders, Traidcraft has been at the forefront of the development of social accounting techniques, which aim to produce a publicly available annual report of performance against indicators developed in consultation with key stakeholders, and which social auditors then verify.[6] Such an approach may become more important to ATOs as more companies are making claims that they trade ethically.

Fair trade labelling

Fair trade labelling organizations are not involved in trade exchanges themselves, but issue fair trade marks or labels to manufacturers or importers in order to verify that the production and supply of a particular product has met specified fair trade standards. Fair trade labels are issued to a range of commodities for which an international fair trade standard has been developed (these include coffee, tea, banana, rice, sugar, fruit juice, cocoa, honey and sports balls).[7]

A buyer (whether a conventional company or ATO) who sources a product from a registered fair trade producer and trades according to the FLO criteria for that product can apply to use a fair trade label from the relevant FLO National Initiative in the country in which the product is sold. The label shows that the product has been produced and traded according to predefined social, contractual and environmental standards, including the payment of the agreed minimum price as determined by FLO. This price is not only intended to provide a better return to the producer, but includes a 'social premium' to be used by producer groups for social development activities.

Fairtrade labelling emerged in The Netherlands during the late 1980s as coffee prices began to fall in the wake of the collapse of the International Coffee Agreement that had stabilized the coffee market for both producing and consuming countries, but also as fair trade supporters looked for ways of widening the fair trade market for coffee. There were two problems with the fair trade model at the time. First, most people do not buy coffee from shops selling mostly handicrafts and gifts; they go to supermarkets. Second was the poor quality of most coffee that had been sold on a fair trade basis. The solution that Max Havelaar innovated was to develop a Fairtrade standard and label. Any coffee roaster or importer would be able to apply for a Fairtrade label provided that they abided by a set of trading criteria and bought the coffee from a registered producer. Coffee would be both fair trade and good quality.

The idea quickly took root and spread from The Netherlands across Europe and later to the US and the Pacific. The Max Havelaar Fairtrade coffee principles were recognized by the emerging national fair trade initiatives and became the international Fairtrade principles, which later became the foundation of FLO.

Fairtrade standards cover both the conditions of production and the conditions of trade (see Box 2.1). There are different production criteria for small producers organized as co-operatives compared with enterprises employing labour, with the latter focusing on labour standards and structures for worker representation and the former on democratic organizational structures.

Until fairly recently, there were different sets of criteria for different products reflecting when they were developed and the concerns of the movement at the time. Thus, whereas the coffee criteria developed during the late 1980s when coffee prices first began to tumble only had a cursory mention of environmental sustainability, focusing mostly on trading terms, the more recent banana criteria included detailed integrated crop management techniques that are required as minimum Fairtrade criteria, as well as minimum measures to protect the environment. FLO's recent internal review of Fairtrade standards to produce a single set of criteria has adopted the approach used in the banana code, with both minimum criteria and process requirements. Minimum requirements must be met within a specified period after joining the register, whereas process criteria are to be met according to a schedule mutually agreed by FLO and the producer organization. Process requirements include efforts to protect the environment and to improve the effectiveness of co-operatives in representing the needs of all their members.

While Fairtrade labelling was theoretically open to any company, initially only relatively small companies took the opportunity to carry a Fairtrade-labelled product as part of a wider product range. Mainstream brands such as Cadbury and Nestlé have been reluctant to support fair trade activities, although Nestlé adopted its first Fairtrade line in 2005. Nevertheless, the

---

Box 2.1 Components of fair trade labelling standards

Trading criteria include:

- a price that covers the cost of production;
- a social premium for development purposes;
- partial payment in advance to avoid small producer organizations falling into debt;
- contracts that allow long-term production planning;
- long-term trade relations that allow proper planning and sustainable production practices.

Production criteria include:

- for small farmers' co-operatives: a democratic, participative structure;
- for plantations or factories: workers should have decent wages (at least the legal minimum); good housing, where appropriate; minimum health and safety standards; the right to join trade unions; no child or forced labour; minimum environmental requirements.

Source: based on material from the Fairtrade Labelling Organizations International (FLO) website, www.fairtrade.net

---

Fairtrade Foundation has awarded its Fairtrade Mark to 130 products in the UK, many of which are available through mainstream retailers, and in 2002, UK£63 million was spent on goods with the Fairtrade Mark (Fairtrade Foundation, 2003). A significant milestone for fair trade in the UK was when the Co-operative supermarket decided to source all its own-brand chocolate and coffee through fair trade sources in 2002 and 2003, respectively.

The relationship between fair trade and the mainstream market will be raised again later, but first we will consider who fair trade producers are and how the relationship between producers and fair trade buyers is managed and monitored.

Producers, monitoring and management of fair trade relationships

There is a dazzling array of fair trade producers across the world. There is no central record of how many farmers, artisans and labourers are linked to the fair trade market, but they number in the hundreds of thousands. The members of the European Fair Trade Association (EFTA) currently import from over

Table 2.1 *Features of fair trade: Labelling and alternative trade organizations (ATOs) compared*

	Fair trade labelling	ATOs (fair trade brands)
Products	Coffee, tea, cocoa, honey, sugar, orange juice and bananas	A wide variety, including crafts, textiles, fruit preserves, chocolate, nuts and nut butters
Marketing channels	Multiple retailers; natural food retailers; ATO outlets	Mostly ATO outlets (own stores, catalogues, volunteers), although some sold through natural food retailers; some efforts by ATOs to facilitate sales from partners to conventional companies, not necessarily on fair trade terms
Range of producers	Co-operatives; plantations that meet certain criteria	Co-operatives, producer associations, family firms, NGOs
How criteria are set	At product level by a working group from the national labelling initiatives/Fairtrade Labelling Organizations International (FLO) in consultation with the producers and workers concerned	Individual ATOs have principles that guide their operations. Most ATOs subscribe to the common principles of umbrella organizations such as the International Fair Trade Association (IFAT) or the European Fair Trade Association (EFTA)
Price	Pricing formulae linked to conventional markets, except when conventional prices fall below a specified minimum when the minimum is paid. A fair trade premium is added to ensure that fair trade prices are always above conventional prices	Prices are set to give adequate remuneration to producers (at least covering costs of production)
Nature of criteria	Terms of trading and production conditions (including working conditions for private operators)	Terms of trading, trade development and choice of producer organization
Auditing	Producers and buyers are regularly monitored against FLO criteria	Internal monitoring by ATOs as part of ongoing process

700 producer organizations in Africa, Asia and Latin America (EFTA, 2001), and in May 2004 there were 389 groups certified by FLO (FLO, 2005).

The simplest way to categorize fair trade producers is according to the products they produce and the way in which they are organized. A major categorization is between food and non-food products. Within the food category one can distinguish between primary commodities and processed products made by the producer groups – for example, between coffee and chutneys.[8] In the non-food category there is considerable diversity of products (and, hence, kinds of producer), which are largely handmade crafts but can also include textiles and garments. The marketing co-operative tends to be the most common form of organization amongst both food and non-food fair trade producers. But a co-operative can take a number of forms, from a formally registered hierarchical structure with over 50,000 members (like East African coffee co-operatives), to a loosely structured group of women coming together to market embroidered cloth. Some fair trade producers, however, are organized as companies – for example, the plantations supplying fair trade bananas in Ghana and businesses selling to Traidcraft, such as Agrocel in India and Dezign Inc in Zimbabwe. Privately owned companies producing for the fair trade movement tend to have explicit social missions, and fair trade buyers generally insist on special arrangements for worker participation, especially in relation to decisions over the fair trade premium. Overall however, most fair trade organizations tend to work with groups rather than private individuals.

Producers have become involved in fair trade relationships in a variety of ways; but often, chance plays an important role – a link was established with a fair trade buyer at the right moment. There are considerably more producers deserving of fair trade links than can get access to the fair trade market. For many fair trade organizations, the issue of whether to buy more from existing partners or to develop new relationships is an eternal dilemma, especially when they often buy only a small proportion of any one producer's output.

While they have guiding principles that govern the selection of producers and overall objectives for trading relationships, ATOs do not undertake formal certification against set criteria for most of their products. However, IFAT is introducing a system of self-audit for producers, the results of which will be available to other producers regionally and up and down the supply chain. The challenge for fair trade is to develop cost-effective systems for monitoring many producers, each of which is trading in relatively low volumes of goods, while at the same time ensuring that poor and disadvantaged groups can take advantage of the opportunities offered by fair trade.

In contrast, to ensure that producer groups are delivering benefits to their members, FLO – the custodian of fair trade labelling standards – certifies both buyers and producers. Producers are usually visited once a year by a fair trade monitor who assesses the group and its production processes against the FLO standard, including the way in which the producer group determines how the fair trade premium is used. Where the fair trade-registered producer is a company that hires employees, the fair trade committee is particularly important. This is not only to ensure that the expenditure of the premium is determined by the workers, but as a vehicle to empower workers and enhance their voice in the workplace.

Fair trade and development

Fair trade is rooted in development. The web pages of fair trade organizations are full of encouraging stories about how fair trade relationships have helped producers make a living and provide for their families. These are often contrasted with the poor returns from conventional market outlets or exploitation by intermediaries, characterized as 'coyotes' and 'charlatans'. Interestingly, until fairly recently there had been very few studies undertaken to assess the development impact of fair trade on individual producers and their families, the co-operatives and groups marketing their produce or on the wider community. During the late 1990s, a group of researchers set out to review assessments undertaken by fair trade organizations to see 'who benefits', but were surprised to find out that very few impact studies actually existed and most evidence has been 'anecdotal, without rigorous baseline information or downstream data that demonstrate how this leads to development benefits at the local level' (Robins et al, 1999, p2). This is not to say that the anecdotal evidence of positive benefits does not reflect reality; rather, it is to stress its lack of rigour and the lost opportunity to learn from past experience. During the last few years there has been a surge in studies that have documented the impacts of fair trade on producers that offer some important lessons for fair traders. Some studies focus on crafts and so are not considered here in detail, although some issues raised in studies such as Hopkins's (2000) assessment of producers for Oxfam are also relevant to food producers (such as differential benefits and the potential to create dependency).

Fair trade in food commodities

Impact assessments of fair trade in food have largely focused on the effects on the producer organization and prices received by farmers. One exception is a series of studies that have examined the implications for livelihoods of producers, taking the analysis beyond the narrow commodity focus of many discussions (NRET, 1998, and Nelson et al, 2002, henceforth referred to as the NRET studies). Another exception is a study of the impact of fair trade on the Day Chocolate Company that not only considers impacts on producers in the Kuapa Kokoo co-operative, but also impacts in the consuming country (Ronchi, 2002).

The Natural Resources and Ethical Trade Programme (NRET) studies suggest that direct financial benefits of fair trade tend to be restricted to members of the producer organizations, usually male heads of household, and are not spread to the wider community. Fair trade has strengthened some co-operatives and, in some cases, has offered them a lifeline; but there is potential for greater capacity-building. Higher prices and fair trade premiums may lead to investment in physical capital, which can be bolstered by donor grants, which may be more readily accessible due to links with fair trade organizations (the honey pot syndrome). Generally, there was limited explicit consideration of environmental sustainability in the schemes assessed. The sustainable livelihoods lens helps one to suggest additional inputs, building up the export capacity of specified groups of producers, which may ensure that the development impact is maximized. For example, in the Brazil nut case one could take steps to ensure that the shellers, as well as the Brazil nut concession-holders, benefit from fair trade (Nelson et al, 2002).

Ronchi's (2002) study was undertaken as part of a process to develop a sustainable monitoring and evaluation framework for both the cocoa co-operative and the chocolate trading company that could be used to improve the quality of their work. Its approach was strongly participatory, beginning by developing objectives with the key stakeholders. At the cocoa co-operative level, the study highlighted success in generating income for farmers from fair trade premia, but also as a result of Kuapa Kokoo's profitable operations, which have been supported through fair trading links, and the advice and technical support from Day Chocolate and Twin. Areas to improve include female representation and the administrative skills of primary co-operative representatives. In terms of fair trade chocolate retailing, Day Chocolate has reached consumers who were previously unaware of the concept of fair trade; but there are many opportunities for increased sales that have not yet been tapped.

General lessons

There is increasing recognition that price is not the only benefit of fair trade. In general, the main contribution of fair trade to many of the groups studied is the development of capacity on the part of the producer group (Tallontire, 2000; Oxford Policy Management, 2000; Ronchi, 2002). For example, a link with fair trade organizations can be a major factor in enabling cocoa and coffee exporters to develop business and technical skills, including assisting co-operatives in providing transparent market information to their members. For many fair trade practitioners, the ideal fair trade relationship is an apprenticeship in which producers learn skills that they can also use when exporting to conventional markets, especially in the North. However, it may be more accurate to say that successful fair trade benefits small producers in poor countries as opposed to saying that fair trade benefits the poor *per se* as 'there are actual and potential negative impacts, particularly for those unable or unwilling to participate, and initiatives are weak in targeting certain disadvantaged groups' (NRET, 1998, p36). A number of successful fair trade projects have benefited some (normally wealthier, male) producers, rather than achieving equitable distribution of benefits throughout the community. The single-commodity focus of fair trade means that it does not take into account the wide range of livelihood strategies undertaken by a household, potentially undermining sustainability; for a producer household, the commodity sold on a fair trade basis may be just one part of a diverse production system (Nelson et al, 2002). However, the benefits of other approaches linking producers to Northern markets, such as contract farming, are likely to be even more geographically and socially concentrated. Moreover, there appears to be increasing pressure within the fair trade movement for gender issues to be addressed more thoroughly, as demonstrated by the clearer specification of gender issues in the revised fair trade labelling criteria.

Fair trade business, now and in the future

Fair trade has commercial as well as developmental objectives, which presents challenges for companies engaged in ethical sourcing. Since the development of the fair trade label, the issue of mainstreaming fair trade has edged to the centre of fair trade debates. Yet, how can fair trade reach into the pockets of more consumers and extend the benefits of more equitable trading relationships to more producers?

While there is an increasing awareness of the fair trade concept and, an ever expanding range of Fairtrade marked goods,[9] fair trade products still represent a fraction of the total market. For example, while Cafédirect is now the sixth largest coffee brand in the UK, with a turnover of UK£10 million in 2002, it has captured only 1.9 per cent of the instant coffee market, in contrast to the 60 per cent share of Nescafé. Many fair trade practitioners, especially those that have focused on offering a mainstream product, such as Cafédirect and Day Chocolate Company, argue that this niche can be expanded. For example, in the roast and ground market, fair trade-labelled coffee brands made up 14 per cent of the market (Fairtrade Foundation, 2003).

The decision of UK supermarket the Co-operative Group to source all of its cocoa and, subsequently, coffee from fair trade producers is a landmark in the mainstreaming of fair trade.[10] It is of major importance to the producers listed on fair trade registers and marks a significant moment in the evolution of fair trade business models.

There is an increasing diversity of approaches to fair trade. Even amongst ATOs there are different business management and ownership models that have varying implications for both producer and consumer relations. Most simply, an ATO can be a private company, limited by guarantee, often owned in trust by a linked NGO (Oxfam Trading). Traidcraft is a public limited company with publicly issued shares, a route that Cafédirect has recently decided to take, having been initially established as a partnership of four ATOs. This offers the potential for greater responsiveness to fair trade supporters, especially in the country of incorporation, usually the consuming country. For Cafédirect, communication with producers has been through regular producer conferences, which have been used to guide strategy. North–South partnerships have been taken a stage further in the Day Chocolate Company. The company is one-third owned by Kuapa Kokoo, which is represented on the board of directors (Twin Trading owns 52 per cent and the Body Shop owns 14 per cent). One of four annual board meetings is held in Ghana (Ronchi, 2002).

Once one considers companies using fair trade labels, the range of business models and corporate philosophies involved in fair trade increases further. Some companies with a fair trade label on their products have an explicit commitment to ethical sourcing and purchasing practices. However, by and large, fair trade as practised by mainstream companies with a fair trade label tends not to affect their core business operations; rather, it is a marketing tool, 'one of the tools in the CSR armament rather than a basis for doing business' (Young, 2003). This may not matter if producers are benefiting from a long-term contract and a price significantly above market levels. However, this ignores the importance of the additional services that links with fair trade organizations bring to small-scale producers, such as capacity-building and facilitating market access. As attempts to mainstream fair trade accelerate, it becomes increasingly important to consider what may be lost from the fair trade model when conventional business gets involved.

Conclusions

Fair trade has evolved rapidly over the past half century, with its profile and sales expanding markedly over the past ten years. Fair trade has become increasingly professional and able to engage in mainstream markets. The key principle of empowering disadvantaged producers through better terms of trade, including a fair price, remains a central tenet of fair trade. However, there is a constant but potentially creative tension between the development and trade objectives of fair trade. It is likely that this tension will become more significant as attempts to mainstream fair trade continue.

ATOs will need to decide whether to focus on competing in the mainstream market with conventional companies with labelled products (where fair trade labels are one of an increasing number of sustainability labels) or to adopt other approaches to enabling producers to gain greater benefits from trade. These approaches may include an exclusive focus on trade development and capacity-building, rather than engaging directly in trade or advocacy and lobbying. The danger of an exclusively mainstream approach is that only those producers who have reached a certain level of organization, export capability and quality will be able to enter a fair trade market; as a result, fair trade could become a path for only an elite set of producers. It is important that the trade development and market access elements of fair trade are not lost. Nevertheless, there is still an important role for fair trade labels and standards, especially since they have a distinctive focus on economic, as well as social and environmental criteria.

Notes

1   Fair trade sales reported for 2002 by the Fair Trade Federation covering the US, Canada, Australia, New Zealand and Japan were US$250 million (Fair Trade Federation, 2003).

2   Between 2001 and 2003, worldwide sales of fair trade-labelled products grew by 21.2 per cent (FLO, www.fairtrade.net/sites/impact/facts.htm, accessed 30 January 2004).

3   These are based largely in Western Europe, but also in the US and Japan, with an office in Mexico.

4   Formerly known as the International Federation for Alternative Trade, hence the acronym; see www.ifat.org.

5   In the academic literature on fair trade, US scholars reflect greater idealism compared with the pragmatism of European writers, although as Raynolds notes in this volume (see Chapter 3), ideas and practice are starting to change in the US.

6   International fair trade criteria are also being considered for garments and textiles; the Swiss national initiative has a Max Havelaar fair trade standard for cut flowers and roses with a Fairtrade Mark, which are now available in the UK.

7   This is a different process from social auditing against a code of practice, discussed elsewhere in this book (Chapter 8); it is a more open-ended process of consultation with stakeholders about a company's overall performance.

8   However, processed goods such as chocolate and muesli bars tend to be manufactured in Europe by others.

9   By 2004, 140 fair trade-labelled products were on the market in the UK, and retail sales of goods with a Fairtrade Mark reached a retail value of UK£63 million, a 90 per cent increase from 2000 (Fairtrade Foundation, 2003).

10   After switching to fair trade cocoa sources in November 2002, Co-op brand chocolate sales increased by 21 per cent. The effect of the Co-op buying only fair trade coffee will increase nationwide sales of Fairtrade Mark coffee by 15 per cent (Co-operative Group Advertisement Promotion, *Guardian Weekend*, 15 November 2002).

References

EFTA (European Fair Trade Association) (2001) 'Lets go fair!', *Fair Trade Yearbook*, EFTA, www.eftafairtrade.org/pdf/YRB2001ch02_EN.pdf

Fair Trade Federation (2003) *Fair Trade Trends in the US, Canada and Pacific Rim*, Fair Trade Federation, www.fairtradefederation.com/2003_trends_report.pdf

Fairtrade Foundation (2003) *Highlights of 2002*, Fairtrade Foundation, www.fairtrade.org.uk

FLO (Fairtrade Labelling Organizations International) (2005) *Facts and Figures: Impact*, FLO, www.fairtrade.net/sites/impact/facts.html

Hopkins, R. (2000) *Impact Assessment Study of Oxfam Fair Trade*, Oxfam, Oxford

Littrell, M. A. and Dickson, M. A. (1999) *Social Responsibility in the Global Market: Fair Trade of Cultural Products*, Sage, Thousand Oaks, London and New Delhi

Nelson, V., Tallontire, A. and Collinson, C. (2002) 'Assessing the benefits of ethical trade schemes in cocoa (Ecuador) and brazil nuts (Peru) for forest dependent people and their livelihoods', *International Forestry Review*, vol 4, no 2, pp99–109

NRET (Natural Resources and Ethical Trade Programme) (1998) 'Ethical trade and sustainable rural livelihoods', in Carney, D. (ed) *Sustainable Rural Livelihoods: What Contribution Can We Make?*, Department for International Development, London

Oxford Policy Management (2000) *Fair Trade: Overview, Impact, Challenges. Study to Inform DFID's Support to Fair Trade*, Oxford Policy Management, Oxford

Robins, N., Roberts, S. and Abbot, J. (1999) *Who Benefits? A Social Assessment of Environmentally-driven Trade*, International Institute for Environment and Development, London

Ronchi, L. (2002) *Monitoring Impact of Fairtrade Initiatives: A Case Study of Kuapa Kokoo and the Day Chocolate Company*, Twin, London

Tallontire, A. (2000) 'Partnerships in fair trade: Reflections from a case study of Cafédirect', *Development in Practice*, vol 10, no 2, pp166–178

Young, G. (2003) 'Fair trade's influential past and the challenges of its future', Paper presented at Fair Trade: An Asset for Development conference organized by the King Badouin Foundation, Brussels, 28 May 2003, available at www.kbs-frb.be/files/db/en/PUB%5F1337%5FFair%5FTrade.pdf

3

Organic and Fair Trade Movements in Global Food Networks

Laura Raynolds

Introduction[1]

The international organic agriculture and fair trade movements represent important challenges to the ecologically and socially destructive relations that characterize the global food system. Both movements critique conventional agricultural production and consumption patterns and seek to create a more sustainable world food system. The international organic movement focuses on re-embedding crop and livestock production in 'natural processes', encouraging trade in agricultural commodities produced under certified organic conditions and processed goods derived from these commodities. For its part, the fair trade movement fosters the re-embedding of international commodity production and distribution in 'equitable social relations', developing a more stable and advantageous system of trade for goods produced under favourable social and environmental conditions.

The international trade in organic and fair trade products represents a relatively minor share of the global market, but this trade is growing rapidly. The world market for organic products is worth about US$20 billion, with sales growing at over 20 per cent per year in many countries (Sahota, 2004). Although much of the demand for organic foods in the global North is met by local production and the North–North organic trade, the fastest growth is in imports from countries of the global South (Raynolds, 2004). Fair trade networks, which by definition link Southern producers with Northern consumers, are younger and less well developed than organic networks (Raynolds, 2002). Fair trade sales are worth about US$1 billion per year, yet are expanding even more rapidly than those in organics, with annual growth rates averaging 30 per cent across Northern markets (FLO, 2005d).

This chapter analyses the recent boom in North–South organic and fair trade food networks, and the similarities and differences between these two movements. I argue from a theoretical and empirical basis that what makes fair trade a more effective oppositional movement is that it moves beyond the realm of production to challenge unequal trade relations. By demystifying global relations of exchange and challenging market competitiveness based solely on price, the fair trade movement creates a progressive opening for bridging the widening North–South divide, and for wresting control of the food system away from transnational corporations infamous for their socially and environmentally destructive business practices.

Background

Initiatives that promote trade in commodities which claim to have been produced under more socially or environmentally sustainable conditions have proliferated over recent years (Blowfield, 1999; Barrientos, 2000; Gereffi et al, 2001). Some of these efforts represent little more than corporate public relations campaigns aimed at making their products look 'cleaner' and 'greener' than comparable products sold by their competitors. But many ethical sourcing initiatives involve rigorous oversight by third-party groups that certify compliance with established social and/or environmental production standards. Such initiatives exist in apparel, footwear, textiles, timber, flowers, marine products and other sectors. The fair trade and international organic movements represent two of the most powerful initiatives in the global food sector.

Voluntary certification and labelling efforts convey information to consumers about the particular social and environmental conditions under which commodities are produced. In international arenas, these initiatives address Northern consumers' rising concern over the global implications of their purchasing practices. These initiatives are particularly powerful in the food sector, where production conditions are linked to the health and safety of products for consumers. In purchasing certified commodities, consumers grant market shares and often premium prices to participating distributors and producers.

The international organic and fair trade movements have both created impressive markets for labelled commodities. Yet, I suggest that their true significance lies not in their market shares, but in the challenge they raise to conventional market principles. Conventional capitalist markets are guided by prices that understate the full ecological and social costs of production and thus encourage the degradation of environmental and human resources, particularly in the global South. The international organic and fair trade movements make visible the ecological and social relations embedded within a commodity and ask that Northern consumers shoulder a greater share of true production costs. These efforts require the creation of tighter links between Southern producers and Northern consumers, as well as the introduction of alternative products. The organic movement goes further in specifying the ecological conditions and costs of production; the fair trade movement, in turn, goes further in detailing the social conditions and costs of production. Both clearly question the practices and prices guiding conventional global food production and trade.

Although commendable, acknowledging the production conditions of a food commodity informs us of only a portion of the relations embodied in an item found on our supermarket shelves. Where organic certification is silent about conditions beyond the point of production, fair trade initiatives seek to make transparent the relations under which commodities are exchanged. By demystifying global trade and creating more equitable relations of exchange, the fair trade movement goes further in challenging conventional market practices.[2] I argue that, for theoretical and empirical reasons, this social re-embedding of exchange, as well as production relations, is essential for countering ecologically and socially destructive practices in the current global food system.[3]

The international organic agriculture and fair trade movements

The organic and fair trade movements represent important ethical sourcing efforts that both challenge existing production and consumption patterns and seek to create a more sustainable world food system. These initiatives originate in the global North and are fuelled by mounting concern that our modern state and corporate institutions are unable to guarantee the socially and environmentally sound production of consumer goods. The organic and fair trade movements are each buttressed by a strong transnational non-governmental organization (NGO) which links Southern producers to Northern consumers via voluntary commodity certification (see the discussion of IFOAM and FLO in the following section). Products which meet a set of production and/or trade criteria can be labelled as organic and/or fair trade, with compliance verified by a third party. Although organic and fair trade certification systems operate independently, particular products can be (and many are) certified as both organic and fairly traded.

International organic organization and standards

The international organic agriculture movement grows out of diverse initiatives in the US and Europe that criticize the unsustainable character of industrial agriculture and the unhealthy nature of agro-industrial foods. These initiatives seek to create a healthier and more sustainable food system by re-embedding crop and livestock production in 'organic' or 'ecological' processes. While there is no one definition of organic agriculture, there is general agreement that this represents a system of farm management based on natural methods of enhancing soil fertility and resisting disease, rejection of synthetic fertilizers and pesticides, and minimization of damage to the environment. Over the past 30 years organic standards and certification procedures have been institutionalized nationally via legislation in major North American, European and other markets; these regulations have been extended internationally through the multilateral efforts of the Codex Alimentarius Commission (Raynolds, 2004).

Established in 1972, the International Federation of Organic Agriculture Movements (IFOAM) has worked to promote the world market for certified organic commodities and to harmonize organic standards and procedures worldwide. With 75 members from 100 countries, IFOAM pursues a 'holistic approach to the development of organic farming systems, including the maintenance of a sustainable environment and respect for the needs of humanity' (IFOAM, 2005a). The organization has established a set of detailed agro-ecological requirements that must be satisfied for products to be certified as organic. These organic production standards are now codified, with little variation, in national and international organic certification and labelling legislation. While IFOAM has recently introduced basic human rights issues in its organic standards, these are not mirrored in national or international law. Organic certification procedures are similar worldwide and largely follow IFOAM verification, auditing and documentation standards, even though only some organic certifiers are IFOAM accredited. Certification costs are borne by producers. Table 3.1 outlines key organic standards and certification procedures.

Fair trade organization and standards

The fair trade movement has grown largely out of European initiatives seeking to transform North–South trade from a vehicle of exploitation to one of sustainable development (Renard, 2003). Alternative trade organizations import fair trade products that are sold to socially conscious consumers in Europe and, to a lesser degree, in North America. By building consumer–producer solidarity, fair trade seeks to re-embed the production and marketing of exports from countries of the South in more equitable social relations (Raynolds, 2002).

Fair trade has grown rapidly over recent years, largely as a result of labelling initiatives that have made certified commodities available in mainstream outlets. Three fair trade labels, TransFair, Max Havelaar and Fairtrade Mark, were introduced in Europe and later extended to the US, Canada and Japan. Fair trade certification standards and procedures are harmonized under the auspices of the Fairtrade Labelling Organizations International (FLO). Alhough fair trade labelling initially focused only on coffee, FLO currently provides certification guidelines for 12 commodities (FLO, 2005a). Fair trade standards have no parallel or protection in law, as exists for organics. As summarized in Table 3.1, fair trade requirements specify strict social standards and only democratic groups of small growers or plantations where workers are represented by independent unions can be FLO certified. FLO environmental criteria are lower than those in organics; but many producers go on to get organic (as well as fair trade) certification. Importantly, fair trade standards focus on trade as well as production conditions. Fair trade buyers

Table 3.1 *Organic and fair trade standards*

	Organic	*Fair trade*
Certification and monitoring	There is a 12-month conversion period; the initial inspection is followed by annual visits by independent monitors overseen by accredited certifying organizations; certification costs are borne by producers	Acceptance process takes about six months; the initial site visit is followed by annual visits by independent monitors overseen by Fairtrade Labelling Organizations International (FLO). Yearly reports on social and environmental conditions and the use of the fair trade premium are required. Certification fees are being introduced for producers
Type of producers	Unspecified	Requirement that producers are democratically organized associations of small growers or plantations where workers are fully represented by independent democratic groups
Agro-ecological conditions	Requirement that planting material be chemically untreated and not genetically engineered; the basis of fertilization must be organic. Use of synthetic herbicides, fungicides and pesticides is prohibited (with a few exceptions). Land clearing by burning must be regulated	Requirement that attempts be made to protect forests and wildlife habitat, prevent erosion and water pollution, reduce chemical fertilizer and synthetic pesticide use, and compost wastes. Use of herbicides and some specified pesticides prohibited
Labour	Requirement that attempts be made to ensure social justice, protection of indigenous rights, adequate wages and upholding of basic human rights	Requirement to uphold International Labour Organization (ILO) conventions, including rights to association and collective bargaining; freedom from discrimination and unequal pay; no forced or child labour; minimum social and labour conditions; and rights to safe and healthy working conditions
Producer prices and credit	Unspecified	Guaranteed minimum above the world price (includes premium for social/environmental reinvestment) that moves up with the market. Stipulated bonus for organics. Credit advances of 60% of harvest value on request.
Trade relations	Unspecified	Must be as direct as possible and aimed at long-term trading relations.
Logos	Certifying organizations apply own labels that align with certification requirements in major EU and US markets	Carry national fair trade label: Max Havelaar, TransFair or Fairtrade Mark. A joint seal is being introduced

Source: FLO (2005a); IFOAM (2005a)

must guarantee FLO-established minimum prices, pay a social premium, offer producer credit and trade as directly with producers as possible. Fair trade certification procedures are similar, although less bureaucratic, than those in organics. The fair trade system is financed largely through importer licensing fees, although producer certification fees have recently been introduced.

Organic and fair trade consumption in the North

As consumer interest in purchasing alternative foods has grown, organic and fair trade products have moved beyond specialty outlets and are sold increasingly in regular supermarkets. This market expansion has made labels more important in distinguishing alternative products from their conventional counterparts, as well as fostering product trust among increasingly sceptical consumers. In the organic sector, while IFOAM and national movement groups remain important in promoting organic consumption, product guarantees are provided largely through the legal system. In the fair trade sector, FLO, national labelling and alternative trade groups remain central in both promoting fair trade consumption and underpinning the legitimacy of fair trade claims. The organic and fair trade movements both face key challenges from attempts by corporations to appropriate the market shares and price premiums generated in these sectors, and from the dominance of Northern interests and institutions in these ideally democratic networks.

Organic consumption

Organic foods are one of the fastest growing segments of the global food industry, with demand expanding at roughly 20 per cent per year (Sahota, 2004). As outlined in Table 3.2, annual sales of certified commodities are currently valued at about US$20 billion and are concentrated in North America and Europe. The most rapid growth is currently in imports of certified organic foods from the global South (Raynolds, 2004). These imports supplement supplies of organic fresh produce, grains, meat and dairy products, also produced in the North. But the greatest expansion is in tropical organic commodities not grown in the North – most importantly, coffee, tea, cocoa, bananas and citrus fruit – and counter-seasonal shipments of organic fresh fruits and vegetables, including apples, pears, berries, asparagus, courgette and lettuce. There is also significant growth in Northern imports of inputs for organic processed foods, such as baby food, ketchup and fruit drinks. Initially sold only in alternative shops, during recent years mainstream distributors have greatly increased the availability of organic commodities. Throughout the North, supermarket sales are growing the most rapidly, although sales in alternative retail venues are also thriving (Willer and Yussefi, 2004).

The North–South organic trade is fuelled by rising health and food safety and, to a lesser degree, environmental concerns among Northern consumers. While consumers may assume that their organic purchases have progressive

Table 3.2 *Organic and fair trade markets**

Country	Organic markets (US$1 million)	Fair trade markets (US$1 million)
Europe	8950	741
North America	10,150	289
Pacific Rim	400	3
Total	19,500	1033

Note: * Figures are for 2001, except for European fair trade markets.
Source: organic data calculated from Kortbech-Olsen (2003); fair trade data calculated from FLO, 2005d

social implications, the organic trade in many ways re-enforces traditional North–South inequalities. Since Northern demand is largely for certified organic products, Southern exports are governed by highly bureaucratic certification standards and procedures promulgated in the North, and now enshrined in national and international organic legislation. The mainstreaming of organic sales heightens Northern corporate control since supermarkets exert substantial oversight over global sourcing networks (Raynolds, 2004). Because organic principles do not address exchange conditions, they fail to challenge existing North–South inequalities or transnational corporate domination. Within mainstream markets the ecological and social ideals of the organic movement are often subordinated to the profit motives of corporate distributors. Despite these trends, the persistence of movement-oriented consumers and retail venues throughout the North, and the growth of Southern participation in IFOAM and other organic groups, suggests that the potential to assert the more progressive social elements of the organic movement has not been lost.

Fair trade consumption

Valued at roughly US$1 billion, the fair trade market is much smaller than the organic market, but is growing even faster (at 30 per cent per year) due to the addition of new products and markets (FLO, 2005d). As noted in Table 3.2, Europe accounts for 60 per cent of total sales. Fair trade products are sold in 64,000 retail venues across Europe, including supermarkets and specialty outlets (EFTA, 2001). Fair trade markets are much less well developed in North America, with labelled products only introduced in 1997 in Canada and 1999 in the US. Yet, current market growth in North America is greatly outpacing that in Europe. Coffee, the first labelled commodity, forms the core of the international fair trade system, accounting for over half of the value of all sales (Raynolds, 2002). In 2003, European countries sold over 15,000 metric tonnes of roasted fair trade coffee, led by sales in The Netherlands, the UK and Germany (FLO, 2005b). Although sales in Canada remain modest, by 2003 the US had become the world's largest fair trade coffee market, with sales of

3500 tonnes (FLO, 2005b). Fair trade bananas are much less well established, yet appear to also have tremendous growth potential. Fair trade bananas are the second most important fair trade product in Europe and have recently been introduced in the US and Canada with great success (Raynolds, 2003).

The growth of fair trade has been spawned in large measure by the efforts of FLO and other fair trade groups that galvanize the social justice concerns of consumers and encourage their participation in new fair trade networks.[4] One of FLO's (2005a) key operating principles is 'To raise awareness among consumers of the negative effects on producers of international trade so that they exercise their purchasing power positively.'This emphasis on information sharing links consumers more directly to producers, helping to span the North–South divide (Raynolds, 2002). While consumers use these networks to bolster their trust in the social and environmental origins of their food, producers draw on these networks to access expertise and other resources. The fair trade movement and its certification standards and procedures reflect Northern interests; but Southern producer groups are represented in FLO, in alternative trade groups and on the boards of national labelling initiatives. There is a greater emphasis on North–South partnership in fair trade than in organics. Although fair trade networks are not immune to market forces, the fact that certification standards include minimum price guarantees and other trade restrictions limits the opportunities for corporate interests to refashion this progressive movement into a profit-oriented niche marketing scheme.

Organic and fair trade production in the South

Production of organic and fair trade commodities has grown rapidly throughout the global South over the past decade, although quantifying the extent of this growth is hard given the lack of official data. The vast majority of organic and fair trade production in the global South is oriented towards the export market, fuelled (as noted above) by bourgeoning Northern demand and the price premiums typically associated with these specialty products. But recent international and national policy trends have also promoted this move into organic and fair trade production in Latin America, Africa and Asia. The widespread adoption of neo-liberal policies has undermined national support for local producers, increased the costs of chemical inputs and shifted resources towards non-traditional agro-exports (Raynolds, 1997). As national food markets have deteriorated and traditional international markets have become more volatile, Southern producers have been forced to search out new export markets.

Organic production

Over the past decade, organic production has increased tremendously around the world, and currently over 60 countries in the global South export certified organic commodities. As noted in Table 3.3, thousands of enterprises in Asia and Africa export certified organic products such as cotton, tea, coffee and

Table 3.3 *Organic and fair trade production*

Country	Organic production (number of certified enterprises)	Fair trade production (number of certified groups)
Latin America*	142,622	243
Asia**	61,000	70
Africa***	71,000	64
Total	274,622	377

Notes: * Includes Mexico and the Caribbean.
** Includes the former Soviet Union.
*** Includes Turkey.
Source: organic data calculated from Willer and Yuseffi (2004); fair trade data are calculated from FLO (2005c)

cocoa. Yet, Latin America represents the hub of organic production in the global South, with over 50 per cent of the organic enterprises and over 80 per cent of the certified organic land (Willer and Yussefi, 2004). Mexico leads the region in organic exports, with US$140 million in export earnings per year and over 50,000 growers engaged largely in the production of organic coffee. Argentina has the most extensive certified area and is a key exporter of organic grains, fresh fruits and vegetables and meat. In relative terms, organics are the most significant in smaller countries such as the Dominican Republic, which generates 10 per cent of total agro-export earnings from organic foods and is the world's largest supplier of organic bananas and cocoa (Raynolds, 2004).

It is often assumed that organic production will be the domain of small family farms due to its higher labour demands and compatibility with traditional peasant farming practices. But while peasant producers may meet organic production expectations, their produce cannot enter international organic markets unless it is certified. For marginal producers in the South, organic certification is difficult and expensive, absorbing up to 5 per cent of sales returns (Rundgren, 2000). Certification thus raises significant barriers to entry for Southern producers wishing to enter global organic markets. In some commodities such as organic coffee and bananas, small-scale producers predominate; but this is largely because they are also involved in fair trade networks that help to organize and pay for organic certification (Raynolds, 2004). Where organic and fair trade networks do not intertwine, traditional economies of scale in export production and commercialization work to the advantage of larger producers. Transnational corporations are gaining control of strands of the North–South organic trade, as is evidenced, for example, by Dole Food Corporation's expanding presence in the international organic fresh fruit and vegetable trade. Without the strict social standards and restrictions on eligible producers found in fair trade, organic production risks being transformed from a form of alternative agriculture to a segment of the traditional corporate-dominated global food trade.

Fair trade production

There are currently roughly 400 certified fair trade producer organizations, representing hundreds of thousands of small-scale growers, operating in 45 countries of the global South. As noted in Table 3.3, there are numerous registered fair trade groups in Asia and Africa involved in the production of tea, coffee, rice and fresh fruits and juices. Yet, like in organics, the vast majority of certified producer groups (243 in total) are located in Latin America. Over half of the region's fair trade groups produce coffee; the remainder produce honey, fresh fruits and juices, bananas and cocoa (FLO, 2005c). Latin America produces over 84 per cent of all fair trade-certified coffee. Mexico is the world's largest producer, followed by Peru, Colombia and Nicaragua. Roughly half of Latin America's fair trade coffee is also certified organic. Fair trade bananas are not as well established as coffee, but exports from the Dominican Republic, Ecuador and the Windward Islands are growing rapidly (Raynolds, 2003).

The types of Southern partners eligible to participate in fair trade are specified in FLO standards. Initially, fair trade networks were limited to small-scale producers organized into democratic co-operatives. Eligibility has been extended in some commodities to include progressive plantations where workers are represented by independent unions. While fair trade-labelled coffee must come from smallholders, fair trade bananas and tea are produced by both small and large enterprises.[5] In all commodities, the participation of marginal enterprises in fair trade is encouraged by the fact that the costs of maintaining the system are largely paid by importers.

For small-scale producers, the most direct benefits from fair trade come from the higher guaranteed prices. The importance of fair trade price guarantees is clearest in the case of coffee, where the FLO minimum price is currently over twice the world market price. This price floor has meant the difference between survival and bankruptcy for many small-scale coffee growers. In addition to protecting producers from world price slumps, fair trade provides a social premium to be invested in social and environmental projects. Among fair trade coffee groups, this premium supports critical education, health, food self-sufficiency, transportation and coffee improvement projects (Raynolds et al, 2004). For larger enterprises, the fair trade social premium is more important than the price floor in improving worker welfare since the premium is what provides the funding for projects such as purchasing ownership shares and supporting educational, health and housing projects. Yet, for both producers and workers, the most important benefits of fair trade engagement appears, in the long run, to come from the multifaceted informal and formal support provided for organizational capacity-building.

Conclusions

The international organic agriculture and fair trade movements seek to create alternative trade circuits for items produced under more environmentally and socially sustainable conditions which simultaneously parallel and challenge the conventional global food system. These initiatives have created substantial and rapidly growing parallel markets for certified foods. But this is not all that they have done. They have also raised important challenges to conventional capitalist market practices that make invisible and devalue the natural and human resources that go into producing the foods we eat. These initiatives make us more aware of the true ecological and social costs of our consumption practices and insist that we take greater responsibility for ameliorating the negative repercussions of those practices. The true impact of these and other ethical sourcing initiatives lies in their ability to transform the ecologically and socially destructive practices that characterize our conventional global food system.

The international organic movement has already achieved important environmental gains and raised consumer consciousness regarding the hidden ecological and health costs of industrial food production. Efforts are also being made to solidify social concerns within the organic movement. Noting the convergences between the social and ecological values of the organic and fair trade movements, efforts have begun to establish minimum social and environmental standards applicable in both arenas (IFOAM, 2005b). Yet, I contend that the fair trade movement has raised a more fundamental challenge to the conventional food system due to its emphasis on creating more equitable and sustainable relations of exchange, as well as production. I argue that, theoretically, it is in the process of capitalist exchange that commodities become abstracted from their human and natural roots so that price becomes their dominant characteristic. To socially and environmentally re-embed agricultural production would thus appear to require not just alternative products, but alternative marketing links. This chapter suggests that fair trade initiatives have begun to create new networks of exchange that escape the bonds of simple price competition. By building alternative networks of solidarity between food producers and consumers, fair trade initiatives encourage the participation of disadvantaged farmers and thwart the entry of transnational corporations seeking only to profit from lucrative new niche markets. In contrast, in the organic sector, where trade is left to conventional market forces, marginal producers may be excluded, while transnational corporations are permitted to appropriate the value added by organic labels without adhering to the movement's underlying social and environmental values. The lessons learned from the organic and fair trade movements are critical for building a more just and sustainable global food system.

Notes

1   This chapter benefits from collaborative research with Douglas Murray and from funding from the John D. and Catherine T. MacArthur Foundation and the Ford Foundation. I thank the editors for their constructive comments. The views presented here are the responsibility of the author alone.
2   As Marx (1976) explains, it is through capitalist exchange that products are abstracted from their natural and human roots to become impersonal commodities ruled by abstract prices and market forces.
3   For an elaboration of this argument, see Raynolds (2000; 2002; 2004).
4   The recent introduction of own-brand fair trade-labelled goods by mainstream and natural food supermarkets heightens these retailers' interest in also promoting fair trade.
5   A key question being debated in the movement is whether fair trade certification should be available to coffee plantations.

References

Barrientos, S. (2000) 'Globalization and ethical trade: Assessing the implications for development', *Journal of International Development*, vol 12, no 4, pp559–570
Blowfield, M. (1999) 'Ethical trade: A review of developments and issues', *Third World Quarterly*, vol 20, no 4, pp753–770
EFTA (European Fair Trade Association) (2001) *Fair Trade in Europe*, EFTA, Maastricht
FLO (Fairtrade Labelling Organizations International) (2005a) 'About FLO', FLO, www.fairtrade.net/sites/aboutflo
FLO (2005b) 'Coffee markets', FLO, www.fairtrade.net/sites/products/coffee/market
FLO (2005c) 'Product partners', FLO, www.fairtrade.net/sites/products/partners
FLO (2005d) 'Annual Report 2004/2005', FLO, www.fairtrade.net/sites/news/FLO_AR_2004_05.pdf
Gereffi, G., Garcia-Johnson, R. and Sasser, E. (2001) 'The NGO–industrial complex', *Foreign Policy*, July/August, pp56–65
IFOAM (International Federation of Organic Agriculture Movements) (2005a) 'About IFOAM', IFOAM, www.ifoam.org
IFOAM (2005b) 'Organic agriculture and fair trade', IFOAM, www.ifoam.org
Kortbech-Olsen, R. (2003) 'Market', in Yussefi, M. and Willer, H. (eds) *The World of Organic Agriculture*, IFOAM, www.ifoam.org
Marx, K. (1976, originally published in 1867) *Capital: Volume 1*, Vintage Books, New York
Raynolds, L. (1997) 'Restructuring national agriculture, agro-food trade, and agrarian livelihood in the Caribbean', in Goodman, D. and Watts, M. (eds) *Globalising Food: Agrarian Questions and Global Restructuring*, Routledge, London, pp119–131
Raynolds, L. (2000) 'Re-embedding global agriculture: The international organic and fair trade movements', *Journal of Agriculture and Human Values*, vol 17, pp297–309
Raynolds, L. (2002) 'Consumer/producer links in fair trade coffee networks', *Sociologia Ruralis*, vol 42, no 4, pp404–424
Raynolds, L. (2003) 'The global banana trade', in Moberg, M. and Striffler, S. (eds) *Banana Wars: Power, Production, and History in the Americas*, Duke University Press, Durham, NC, pp23–47

Raynolds, L. (2004) 'The globalization of organic agro-food networks', *World Development*, vol 32, pp725–743

Raynolds, L., Murray, D. and Taylor, P. (2004) 'Fair trade coffee: Building producer capacity via global networks', *Journal of International Development*, vol 16, pp1109–1121

Renard, M.-C. (2003) 'Fair trade: Quality, market and conventions', *Journal of Rural Studies*, vol 19, pp87–96

Rundgren, G. (2000) 'Challenge for developing countries to establish an organic guarantee system', in IFOAM (eds) *Year 2000: The Development of Markets and the Quality of Organic Products Proceedings of the 6th International IFOAM Trade Conference on Organic Products,* IFOAM, Germany, p62

Sahota, A. (2004) 'Overview of the global market for organic food and drink', in Willer, H. and Yussefi, M. (eds) *The World of Organic Agriculture: Statistics and Emerging Trends*, IFOAM, Bonn

Willer, H. and Yussefi, M. (2004) *The World of Organic Agriculture: Statistics and Emerging Trends*, IFOAM, Bonn

4

Corporate Social Responsibility from a Supermarket Perspective: Approach of the Co-operative Group

David Croft

Introduction

Consumer groups, campaigning organizations and stock market investors are increasingly targeting food manufacturers and supermarkets, scrutinizing their activities within the food supply chain. Media coverage draws consumer attention to the sources of their daily shopping and, for the first time, is making the public aware of the way in which the food industry and the food they buy impacts upon communities, farmers, workers and the environment. There is a growing consumer appreciation of the new global food market and its truly worldwide social and environmental impact.

This scrutiny can lead to public criticism and can pose a risk to corporate reputations, and many organizations have developed programmes to manage ethical aspects of their supply chains. These can have various aims, considering such issues as environmental practice, workplace and social standards, or the source of raw materials. For companies, the drive for ethical business also relates to factors ranging from brand protection and business risk management, to stakeholder interest and legislative compliance. Whether motivated by altruistic or commercial reasons, there is no doubt that retailers and brands can exercise significant influence upon their supply chains through their commercial activities. As a result, consumers increasingly expect businesses to effectively and responsibly manage their supply chains. Indeed, as awareness of issues increases though greater media exposure, or as that exposure focuses upon worker exploitation or environmental violations, this pressure will grow, leading more organizations to develop ways of tackling ethical concerns.

This chapter considers how these pressures have introduced new approaches to product sourcing in the supply chain. It specifically focuses on the ways in which organizations and the UK Co-operative Group, in particular, have considered ethical trade and fair trade within their supply chain as a way of fostering more equitable trading practices and workplace conditions. The chapter reviews the differences between these approaches and the impacts they have upon the communities and workers at the heart of product supply.

Developing consumer interest in ethically traded products through fair trade

Growing consumer awareness of fair trade has increased interest in product supply chain issues and the labour conditions associated with the production of everyday items for Northern consumers. While many consumers may feel powerless to address these issues themselves, viewing them as remote or potentially inherent to the products they buy, others are turning to fair trade as a way of ensuring that the products they buy are produced in a responsible manner. However, fair trade alone is not responsible for the increasing consumer awareness of workplace practices. Media coverage, together with the activities of campaigning development organizations and government activity in the UK and the US, has identified areas where improvements in conditions are necessary.

The growth in consumer awareness acquired through extensive media coverage is beginning to lead to greater consumer expectations that corporate business considers ethical trade and workplace standards within its broad corporate social responsibility strategy and increasingly adopts and sells ethically traded products where these exist. The broad range of companies addressing ethical trade issues bears witness to this, with the growing membership of the Ethical Trading Initiative (ETI) in the UK, and similar activities in Europe and the US (ETI, 2004).

Almost inevitably, this is leading to an increased number of products with an ethical pedigree. Some of these relate to the living standards of farmers or to the labour conditions of workers involved in their production. Others relate to broader areas of social responsibility, such as the environment, while some claim to combine both. Examples of the latter include Kraft Jacob-Suchard's adoption of Rainforest Alliance-certified sources of coffee, in which 5 million lbs of coffee are sourced through a direct link to growers under a scheme that considers environmental performance and local community education (Rainforest Alliance, 2004). One of the more progressive approaches is the Thandi project in South Africa, developed by wine producers including Dr Paul Clüver, Capespan and local farmers, with the support of UK retailers such as the Co-operative Group, J. Sainsbury and Tesco. The project is developing fair trade products, including oranges, grapes and wine, with improved returns to the farm workers and the establishment of worker committees. Fundamental to the scheme has been the progressive training and promotion of black managers and winemakers, community development and the transfer of farm ownership to co-operatives of local farmers, an approach that has won the recognition of the South African government (T. M. Mbeki, pers com, 2004).

However, such fundamental changes to the supply chain, as demonstrated by the Thandi project, remain rare, with most ethical credentials based upon site assessment activities, or certification of workplace standards through approaches such as Social Accountability 8000 (Social Accountability International, 1997). The majority of site assessment activities do not provide external accreditation, nor are consumers aware that the products they purchase are ethically produced. Overall, most corporate engagement in ethical trade remains largely focused on brand protection, reputation management and investor satisfaction. The influence of such approaches on broader supply chains or on the living and working standards of the many smallholder producers involved in the global food industry remains to be seen. They rarely, if at all, demonstrate the kind of development that is seen through approaches such as fair trade.

The Co-operative Group

The Co-operative Group is the world's largest consumer-owned co-operative. Originally established in 1863 to provide wholesome, safe and unadulterated foods for its consumer members, the Co-op has grown significantly from the original shop. Its business portfolio now includes an expanding retail estate of over 1700 food and non-food stores, a major bank and insurer, the largest farming business in the UK, the largest UK funeral provider, a pharmacy chain and the largest independent travel agency. While these originally developed to serve members that were either individual consumers or other co-operative societies, they now serve a broader consumer base, and each incorporates the values and principles of the co-operative movement in their activities.

Although the Co-operative Group's current activities are somewhat different from those of the 1860s, the original values remain relevant to consumers and members of the Co-op. Delivering these values, through various corporate policies and practices, supports a broad approach to corporate social responsibility (CSR). Often, the Co-op's activities are viewed as groundbreaking, either because of the processes underpinning them or through the specific activities involved, with peers recognizing this through a variety of awards to the Co-operative Group or its divisions. For example, the Co-operative Bank was declared the 2003 winner of the European Sustainability Reporting Awards, while Co-operative Retail has been recognized as the leading supermarket supporter of fair trade and a winner of the Green Apple award in the environmental sphere for its work on improved pesticide management.

The Co-op's corporate governance monitors how these values and principles are applied in the group's activities. The Values and Principles Committee oversees this function and reports directly to the Co-operative Group's board, whose constitution includes democratically elected consumer members sitting alongside corporate representation from co-operative societies. As a result,

---

Box 4.1 Co-operative Group movement values and principles

The Co-operative Group's values centre on:

- democracy;
- equality;
- self-help;
- solidarity;
- social responsibility.

Its principles are:

- autonomy and independence;
- concern for community;
- education and information;
- cooperation among co-operatives.

Source: Co-operative Group, www.cooponline.coop/about_whatis_values.html

---

the interests of members and the delivery of those defined values remain core elements of business activity and strategy, emphasizing the co-operative difference, an issue highlighted by a 2001 independent review (Co-operative Commission, 2001).

The origins of the Co-op have led the group to develop its CSR strategy from a different perspective than that of most other organizations. This has become obvious through the positions adopted, most clearly in the retail and banking arenas, and with regard to aspects of ethical trade addressing workplace and social standards for producers within the supply chain. In terms of assessing the ethics of their supply chains, the Co-op's strategy involves a programme that moves beyond compliance-based audit approaches, while also developing new solutions to addressing workplace standards through collaborative activity with suppliers and rethinking the structure of the supply chain. For example, through fair trade, the otherwise extensive commodity supply chain for cocoa is considerably shortened through direct links to producers. Whereas the cocoa supply chain can commonly involve up to 15 middlemen, the Co-operative Group's fair trade chocolate chain has only six steps from grower to consumer: the Kuapa Kokoo farmers co-operative in Ghana; the bean processor; the chocolate manufacturer; a UK distributor; the Co-operative Group's shops; and the final consumer. This streamlined supply chain provides more direct contact between producers and consumers, enabling Northern consumers to acquire a deeper understanding of the issues facing farmers, something that the Co-operative Group communicates to its consumers within its fair trade literature (Co-operative Group, 2002).

The Co-op's approach to its supply chain is based upon working with suppliers to improve workplace standards through a shared programme. In doing so, it also seeks to inform and educate suppliers about different ways of working that improve productivity, quality and product standards, while simultaneously improving workplace standards. Collectively, these elements of supplier engagement are now core to the group's approach on ethical trade and have been met with great support from the suppliers involved in such diverse locations as India, Kenya and Colombia, where the group works alongside development organizations to deliver the programme and support its suppliers. In India, for example, the Co-operative Group works closely with development non-governmental organizations (NGOs) and the International Resources for Fairer Trade (IRFT), which help to assess the conditions of suppliers and develop action plans to improve labour conditions for the workers involved.

Potentially more dynamic is the Co-operative Group's adoption of fair trade as a core approach to sourcing commodity-based products from developing countries. This has fundamentally reshaped supply lines for products such as cocoa for chocolate, fruits such as bananas and coffee, while offering much improved returns for smallholder growers.

Ethically traded and fair trade products: The values of each

Ethical trade strategies establish a clear process to confirm that recognized standards relating to the working conditions of suppliers are being met. These standards are generally based upon international labour standards, although they can be integrated within company-generated codes that are applied within a business's supply chain. Not all company codes achieve international standards; but there is growing alignment due to the wide recognition of ethical trade.

In principle, international labour standards, developed via the International Labour Organization (ILO) and governments, should be in place with legislative backing at a national level. However, there are a number of ILO conventions that are not yet adopted by all countries or, where they are adopted, are not applied within relevant national legislation. Equally, local enforcement of labour legislation, or even an understanding of labour standards, is not always present, such that the protection of workers' rights through legislation is not necessarily guaranteed. The trade union movement rightly and forcibly argues that corporate codes to support workplace standards and workers' rights are not a substitute for effective legislation and enforcement (S. Steyne, pers comm, 2003). However, where legislative control does not exist, or where application is patchy, corporate codes of practice can help to ensure the provision of workplace standards that accord with ILO principles. Whether this is originally to protect brand reputation or for more altruistic reasons, commercial relationships can support a drive for reasonable workplace standards. As a result, many organizations are developing ethical trade standards, either individually or through growing numbers of partnerships within the UK ETI and on a wider basis with groups such as the Clean Clothes Campaign (Europe) (ETI, 2004). Typically, those codes address core workplace standards and are summarized within the ETI Base Code (see Chapter 1).

The Co-operative Group, as one of the first members of the ETI, has developed its own code, reflecting closely that of the ETI and the ILO conventions that support it. This code is applied through a range of assessment approaches involving Co-op personnel, external assessors acting on behalf of the group and suppliers' management of the issues that it raises on their own sites. A separate verification approach is also applied in which a sample of assessments is reviewed by a further audit that considers whether the initial review recognized all ethical trade issues on the sites involved.

The group aims to review its supply chain to confirm that ethical trade standards are being met. The ETI plays a combined supporting, coordinating and peer-group review role for all of its members, working through its tripartite membership of the trade union, development and corporate sectors to promote improved workplace standards and learning in the way that they can be achieved. As such, the Co-operative Group is not alone in the UK in addressing these issues, but has been joined by many food and non-food retailers and suppliers with whom working partnerships are being forged. This has led to demonstrable progress in addressing workplace conditions through targeted activities focused on areas of real and perceived risk that are related to poor working conditions. Examples include improved health and safety standards, improved living conditions, application of employment contracts, reduced working hours, avoidance of child labour and increased female supervisory personnel (Co-operative Group, 2003a, 2004).

In some cases, consumers have seen specific products with coordinated marketing promoting their provenance. For example, the Co-operative Group was the first major supermarket to confirm that the supply of its own-brand tea was only from estates that had been assessed and met its ethical trading standards. Based upon work with its supplier, Premier Brands, the relaunch of Co-op Brand 99 Tea included strong on-pack reference to ethical trade assessment, with post-launch sales increasing by almost 50 per cent (Co-operative Group, pers comm, 1999). This growth may not be attributable solely to consumer perceptions of ethical trade or their concerns regarding worker exploitation in the tea industry. The relaunch was also accompanied by packaging that positioned the range as a more 'modern' product offer. Nevertheless, the Co-op and other members of the ETI continue to develop ethical trade within their respective supply chains, reporting upon progress annually and with the ETI publicly reporting on activities (ETI, 2004). It remains clear, though, that while there may have been some review of the wider supply chain, the extent of site-based assessments remains limited by resources and is a small fraction of the enormous web of supermarket suppliers. To date, consideration of the vast numbers of primary producers has been a small part of the process, although the Co-operative Group's approach has addressed some of these. Indeed, the extent of the supply chain raises a concern that a compliance or assessment-based route to improving workplace standards may be impractical. Nevertheless, it appears that the marketing of ethical trade credentials is likely to play a growing part in corporate positioning in the future, rather than simply utilizing ethical trade as a protective measure.

Fair trade

There is a clear distinction between ethically traded and fair trade products. While ethical trade is based upon compliance with established workplace standards through a process of assessment, fair trade is built upon a different concept of the supply chain that enables small producers to access markets and to receive a fair price for their products. It may be overly simplistic to consider the difference in such terms since there are various ways in which large companies, as opposed to alternative trade organizations (ATOs), are developing ethical trade positioning beyond pure compliance with labour standards. However, few, if any, of these currently address trading practice and the question of payment to producers. The premise of fair trade ensures that producers receive a reasonable payment for their crops that might otherwise be subject to massive price fluctuations within international commodity markets. Clearly, small producers have limited or no opportunity to influence these markets. In most cases, the market structure has resulted in payments to growers being far less than their cost of production and, indeed, far less than that which would be necessary to secure basic living standards, such as clean drinking water, medical supplies or education (Co-operative Group, 2002; Oxfam, 2002). Fair trade ensures that the price paid to the grower consistently allows for these amenities, while also including a social premium that supports investment in the local community.

As a result, fair trade argues for a different supply chain that is not operated solely in the interests of global traders, but which supports the needs of small producers. These small producers are usually organized in co-operatives in order to provide sufficient output volume to provide for trade with large organizations with international supply lines. In doing so, fair trade also shortens supply chains, bringing growers into closer contact with the processors of their products and cutting out middlemen who would otherwise take their own cut of the profits out of a chain. These changes can create market access and potentially increase returns to growers. Shortening the supply chain can also facilitate product traceability in a complex commodity market where consolidation of stock is typical and increased raw material traceability is being demanded by legislation (European Commission, 2002).

There is a close relationship between the values of fair trade and those of the Co-operative Group movement already described. In particular, the organization of growers into co-operatives creates a close link to Co-op principles, with the desire of the movement to work in cooperation with other co-operatives. With this synergy in mind, it is not surprising that the Co-operative

Group has developed its range to include fair trade products, a move that has garnered increasing support from its own consumer members who number over 1 million, as well as from shoppers who have demonstrated support for fair trade through increasing sales (Co-operative Group, 2004). It has been argued that fair trade may not provide a long-term solution to the financial crisis facing commodity producers as a result of trends in international trade. But it is also recognized that fair trade can provide a glimmer of hope for the growers and may create commercial leverage through increased fair trade sales that deliver wider change in the future (Oxfam, 2002).

Building consumer awareness and demand for fair trade

The Co-operative Group has set out a clear strategy on fair trade, aiming to lead the UK supermarket sector in its support for fair trade and to bring fair trade into the mainstream, challenging the existing commodity supply chain format and addressing the inequalities that it creates.

Achieving this strategy initially involved the adoption of existing fair trade brands, such as Cafédirect coffee and Divine chocolate; but, since 1999, the Co-op has also focused on the development of the group's own Co-op brand range. Today, entire ranges of Co-op brand products have been converted to fair trade, including, in 2002, all Co-op brand block chocolate (six products) and, in 2003, all Co-op brand coffee (18 products). As high-profile, high-value ranges in a strongly competitive and brand-dominated market, this was a bold strategy to support and develop fair trade in sectors where commodity market trading and world oversupply have led to growers receiving below cost price for their products (Co-operative Group, 2003b). Significant investment in marketing has supported the group's fair trade products and has helped to build sales, the wider fair trade market and broader consumer awareness. These marketing activities included television adverts; in-store activities both in terms of advertising materials and activities with Co-op members surrounding the group's support for Fairtrade fortnight; advertising on the side of delivery vehicles; and promotion at public events, such as the BBC 'Good Food Show' and product sampling at mainline railway stations.

Most significant is the Co-operative Group's adoption of a broad campaign platform for fair trade. At the launch of both the fair trade chocolate and coffee ranges, the Co-op adopted a campaigning stance that encouraged consumers to write to their favourite brands of chocolate and coffee and demand that the manufacturers adopt at least one fair trade line in their ranges (Co-operative Group, 2002, 2003b). The volume sales commanded by these brands would subsequently increase the returns to growers and support the transition of fair trade into the mainstream. Both campaigns were well received by Co-op members, consumers and development organizations; but, to date, most manufacturers of major brands remain reluctant to adopt fair trade sources. Other retailers, however, are developing fair trade lines. For example, J.

Sainsbury, Tesco and Asda have all developed own-label fair trade products, which have collectively boosted the growth of fair trade.

Fair trade products now provide a highly conspicuous element of the Co-op range, with sales increasing from UK£100,000 in 1999 to over UK£24 million in 2004 (Co-operative Group, 2005). The overall UK market for fair trade products during 2002 was in the region of UK£140 million, demonstrating the significance of the Co-operative Group's activity (Fairtrade Foundation, 2005).

Developing and delivering fair trade products for retail

Over time, fair trade products have evolved to become a core element in the Co-operative Group's retail strategy (see Box 4.2). The range has developed rapidly and entailed increasing contact with suppliers and ATOs, which have facilitated the relationship between producers and the Co-operative Group. Indeed, the role of ATOs in developing and supporting the chain has been fundamental to its success, and expanded their organizational function from potentially niche ATOs into significant drivers within the market. Arguably, the size of the Co-operative Group in comparison to larger retailers in the UK necessitated development of these relationships simply by virtue of the need for shared resources. However, the ethos of fair trade is supported by the Co-operative Group's holistic view of supply chain and product development. The increasingly close links with producers, such as the Kuapa Kokoo cocoa co-operative in Ghana, which provides cocoa for the Co-op fair trade chocolate, adds value to the fair trade message and clearly demonstrates the issues faced by growers, while also building links between retailer and producer, and closer ties within the supply chain. This is supported by the producer of the chocolate, the Day Chocolate Company, a joint venture between Kuapa Kokoo, Twin Trading and the Body Shop, which also manages the Divine brand of fair trade chocolate. Their ATO origins and close links to the growers support fair trade and the products' pedigree.

In developing these products, the Co-operative Group worked closely with producers. It also created product ranges that introduced fair trade to mainstream consumer purchasing patterns and opened fair trade to a much wider market, thereby increasing product volumes and providing a greater return to the growers involved. The Co-operative Group estimates that the conversion of the Co-op block chocolate range doubled the UK fair trade chocolate market and will increase Kuapa Kokoo's fair trade cocoa sales by 30 per cent (Co-operative Group, 2002). For Kuapa's approximately 45,000 members, the fair trade cocoa price ensures a stable and guaranteed minimum price, while the additional social premium has provided an estimated 25 new water wells every year supporting almost 90,000 villagers (Co-operative Group, 2002; Kuapa Kokoo, 2003). Such benefits exemplify what Co-op members increasingly expect to see from Co-operative Group activities, and what consumers increasingly expect from fair trade.

Box 4.2 The Co-operative Group's retail fair trade strategy

- *1999*. Strategic decision taken to stock a core range of fair trade products in all Co-operative Group stores, even small convenience shops, making fair trade more widely available. The range included Cafédirect and Percol Fairtrade coffees; Teadirect; and Maya Gold and Divine Fairtrade chocolate. Although commercially challenging, this was essential to build consumer awareness of fair trade, enabling a broader campaign to follow.
- *2000*. In January 2000, the Co-op introduced the UK's first fair trade bananas, leading to a letter of thanks from the Ghanaian government and other supermarkets following suit.

  Co-op fair trade milk chocolate became the first supermarket brand to carry the Fairtrade mark. This marked a strong commitment to fair trade since removing the product from the range would be more significant than simply dropping a branded product. The chocolate, launched with the Day Chocolate Company, also carried the established Divine branding.

  The introduction of Co-op fair trade ground coffee was added to the broad range of branded fair trade lines in an attempt to increase overall fair trade sales.
- *2001*. In the absence of defined fair trade criteria for wines, the Co-operative Group developed activities with Traidcraft to supply a fairly traded wine from the Los Robles co-operative in Chile in February 2001.
- *2002*. The first fair trade mangoes were introduced from Ecuador, working in partnership with Twin Trading.

  A second wine, Semillion, was introduced in partnership with Traidcraft and Los Robles.

  Launch of Co-op Crispy White fair trade chocolate, again with the Day Chocolate Company, meeting a growing market demand for white chocolate.

  Launch of the first fair trade instant coffee granules, positioning fair trade products in the largest instant coffee sector. The product was aimed at the 'mid-sector' in order to avoid competition with established fair trade brands (e.g. Cafédirect, whose freeze-dried coffee brands capture the top end of the premium instant coffee market).

  Launch of the Co-op's fair trade chocolate cake (based on fair trade sugar and chocolate). This marked an important development in composite products and opened a new market for fair trade.

  Conversion of the entire Co-op brand chocolate block range to fair trade, effectively doubling the national fair trade chocolate market valued at UK£3 million, while also developing a consumer campaign platform.

  The world's first fair trade pineapple was introduced from Costa Rica.
- *2003*. Further wine development, including wine boxes from Chile and bottled wines from South Africa.

  Conversion of the entire Co-op brand coffee range into fair trade, thereby applying fair trade to another whole-product category.

  Launch of two fair trade sugars from Malawi.

Source: Co-operative Group (2004)

Ethical trade, developing links to suppliers and long-term sustainable solutions within the supply chain

Developing solutions to ethical trade provides an opportunity to improve workplace conditions through greater dialogue and improved links with suppliers. This changes the emphasis from a top-down perspective of the supply chain, driven by an inspection process, to one that is based upon increased supplier involvement in processes to improve workplace conditions. The Co-operative Group aims for this type of dialogue with its suppliers and has generated supplier partnerships in an increasing number of countries to deliver improved workplace conditions for those in its supply chain.

The process involves a workshop for suppliers in producing countries. To date, this has included India, Kenya, Uganda, Zambia, China, Spain and Colombia. Typically, the workshop introduces the issues of ethical trade and outlines the Co-operative Group's objectives for its supply chain. Most importantly, the workshops aim to provide suppliers with a broader understanding of the business case for ethical trade in order to generate greater buy-in to the ethical trade process and to demonstrate that ethical and commercial incentives are not mutually exclusive. There is increasing evidence to support the view that reasonable workplace conditions can add to commercial performance. For example, improved health and safety may lead to fewer accidents, reduced downtime, and lower staff turnover or absenteeism, all of which combine to create greater stability and reduce training needs, whilst supporting improved productivity and product quality (Co-operative Group, Premier Brands, 2002). The workshops provide information on such benefits through the experience of suppliers who have already developed ethical production systems. This helps to provide suppliers who are new to the process with incentives to engage in ethical trade, such as improved management of their workforce and the potential marketing opportunities available to suppliers who support the aims of their commercial customers.

The workshops also enable suppliers to undertake a self-assessment of their activities against the code of conduct standards and to develop action plans that address areas of concern. These action plans are agreed with the Co-operative Group, and begin to form management systems that embed ethical trade principles in the day-to-day activities on site. The workshops have input from local development organizations, including International Resources for Fairer Trade (IRFT) in India, AfricaNow in Kenya and Fundacion Natura in Colombia, which provide ongoing support to local suppliers. IRFT, for example, has continued to provide input to suppliers following the workshop, with funding for activities and follow-up assessments provided by the Co-operative Group. This approach supports longer-term development of workplace conditions through a process of dialogue, self-development and support. It enables suppliers to generate solutions that are tailored to local needs, and also reduces the need for costly external assessment undertaken by auditors who may lack the local knowledge and connections to assist in developing ethical trade solutions. With suppliers adopting this type of approach within their own systems, whether driven by ethical or commercial incentives, there is a much greater chance that change will become embedded and sustained. In the long term, this is the Co-operative Group's aim, moving the emphasis from a compliance-based policing of codes to an approach that generates improvements from within, supporting international standards, with ownership by suppliers, and working in conjunction with local trade unions and development organizations to the benefit of their workforce.

Ethical supply chain management: Compliance or development?

When companies develop ethical supply chain management, the nature of the process and its goals raises a number of questions. In part, these can be answered by considering a business's motivations for addressing workplace standards in their supply chain. There may be a fundamental goal of brand protection that is over-riding, creating a desire to ensure and, more importantly, demonstrate compliance with defined standards throughout the supply chain. Alternatively, while there may always be an element of reputation management, this can be complemented by aims to improve the working and living conditions of those involved within the supply chain. In this way, brand protection can still be achieved; however, the shift away from a compliance-based approach allows for more far-reaching improvements in workplace practice.

As organizations develop relations with their supply chain, this balance between compliance and development creates a key strategic challenge. The increasing external pressures to demonstrate compliance with a standard, such as avoiding adverse criticism of a brand or pressure from stock market investors, can lead to a desire for a simplistic solution, one that offers protection, and one that may privately aim to shift responsibility for compliance to third-party auditors or the producer. Under these circumstances, evidence and reassurance of compliance, delivered through certification, becomes an attractive prospect.

An approach to certification already exists through the auspices of Social Accountability International and the Social Accountability 8000 standard (Social Accountability International, 1997). Although this has only been achieved by relatively few organizations (655 worldwide by May 2005), the value of an external inspection, referenced to international standards, can clearly help to protect a brand. However, a compliance-based approach may have counter-productive effects. Achieving compliance through inspection and certification of the many thousands of sites involved in supermarket supply chains, particularly to primary producer levels, is an immense task. Consider coffee, for example, sold as major brands or in supermarket own-brands. While there may be one packer, there may be a number of processors creating coffee granules from the roasted bean, as well as a number of roasters. However, the picture dramatically changes in scale when the source of coffee is considered.

For example, in the past, some of the Co-operative Group's coffee came from the Mount Kenya region, where there are almost 150 separate farming co-operatives, each of which has an average of 1300 individual coffee farmers as members. If compliance through the chain is sought, with the aim of trying to address issues such as child labour (as highlighted in the similar supply chain of cocoa), each individual farm may need to be assessed, creating immense questions of resource capacity. This equally presumes that the sources can be identified, a process that can be exceptionally difficult for products such as coffee that are traded through local and international commodity markets and where stocks are aggregated.

Policing standards from the end of the supply chain in this way creates significant operational issues. Operating an audit function to address individual smallholder producers is an immense task, both in terms of the on-site activity and the management processes supporting it. Potentially, samples of smallholders could be audited; but this reduces the rigour of the approach and still necessitates considerable management support to ensure its credibility. A variant on this style has been suggested for cocoa growers in West Africa, where it will involve country licensing schemes operated by governments in order to address concerns over child labour in that sector (Cocoa Verification Design Working Group, 2004).

An audit-driven approach also creates a risk that suppliers will attempt to hide poor labour standards – for example, by maintaining two sets of records, of which one is specifically oriented towards the audit (Verite, 2003). The pressure to demonstrate compliance and remain part of a lucrative supply chain can lead producers to take such steps. Auditors need to be aware of this risk, while buyers at the end of the chain may need to consider the implications of their demands and the expectations of their suppliers.

These pressures, coupled with the resources (financial or technical) required to achieve certification, may also result in small producers not being able to reach the marketplace. The sheer cost of an audit, undertaken by a third-party firm such as an international audit body, may exceed UK£1000 per day, a figure prohibitive for most small producers or those with many sites. However, without such an audit, which is required by some businesses, producers risk losing market access. Businesses at the end of the supply chain simply choose to deselect suppliers who cannot demonstrate standards compliance, leaving the workers involved in potentially poor conditions with reduced opportunity for improvement when business is withdrawn.

An approach that focuses upon standards compliance through auditing stems from corporate culture that emphasizes brand protection. An audit-driven approach is also part of a learning curve that businesses travel through while developing their approach to CSR. In some cases, businesses halt the process at the compliance perspective; but, in others, businesses begin to recognize that in order to make fundamental changes to the way in which their supply chain works they need to adopt an alternative approach. Rather than simply focusing upon compliance with existing workplace standards that may be at the level of lowest common denominator, they aim to develop social and workplace standards with the suppliers and communities involved. This approach involves working with suppliers rather than simply policing their activities. In doing so, these businesses not only adopt a strategy that delivers brand protection and matches espoused corporate values, but also creates positive developments for suppliers and workers in developing countries.

A collaborative approach to working with suppliers challenges the traditional productionist view of managing a supply chain and requires a broader perspective on ethical trade. However, if ethical trade activity, whether it involves fair trade or not, is to have long-term benefits to producers or workers at the base of the supply chain, then those businesses at the apex of the supply chain need to move beyond simple compliance with standards and use their leverage to improve workplace conditions. The Co-operative Group has embarked upon this strategy, thereby supporting its historical values and principles. However, the vast nature of the food supply chain suggests that others, too, need to apply this approach within their own supply chains. This would mark a change from the historical productionist strategies to a new paradigm that recognizes a need to support supply chains more equitably if they are to remain sustainable. While many businesses are beginning to address ethical trade, only a minority appear to be taking that further step. Until more do so, the benefits of compliance-based strategies to the producers and workers supporting supply chains appear limited.

References

Cocoa Verification Design Working Group (2004) *System Overview Draft for Consultation*, Roberts S., National Centre for Business and Sustainability, UK

Co-operative Commission (2001) *The Co-operative Advantage: Creating a Successful Family of Co-operative Businesses*, Report of the Co-operative Commission, UK

Co-operative Group (2002) *The Chocolate Report*, Co-operative Group, www.co-op.co.uk

Co-operative Group (2003a) *Annual Report to the ETI*, Co-operative Group, www.co-op.co.uk

Co-operative Group (2003b) *The Coffee Report*, Co-operative Group, www.co-op.co.uk

Co-operative Group (2004) *Annual Report to the ETI*, Co-operative Group, www.co-op.co.uk

Co-operative Group (2005) Co-operative Group, www.co-op.co.uk/8080/ext_1view point.nsf/0/b901c0c1623dda1780256b7d0040a2a8?OpenDocument

Co-operative Group, Premier Brands (2002) *Supplier Seminar Presentation*, Nairobi, Kenya

ETI (Ethical Trading Initiative) (2004) *Putting Ethics to Work: Annual Report 2003/2004*, ETI, www.ethicaltrade.org/Z/lib/annrep/2004/en/index.shtml

European Commission (2002) *General Food Law Regulation EU 178/2002*, European Commission, Brussels

Fairtrade Foundation (2005) *Fairtrade Shows Massive Public Response to Man-made 'Economic Tsunamis'*, Fairtrade Foundation, London

Kuapa Kokoo (2003) *Annual Report 2003*, Presentation, Annual General Meeting, Kumasi, Kuapa Kokoo, Ghana

Oxfam (2002) *Mugged – Poverty in Your Coffee Cup*, Oxfam, UK

Rainforest Alliance (2004) *Sustainable Coffee Sales Benefit Farms and Workers in Latin America*, Rainforest Alliance, www.rainforest-alliance.org/news/2004/jun30.html

Social Accountability International (1997) *Social Accountability 8000*, Social Accountability International, New York

Verité (2003) *Ethical Trade Training Seminar*, Manchester, UK

5

Ethical Trade: What Does It Mean for Women Workers in African Horticulture?

Sally Smith and Catherine Dolan

Introduction[1]

Much of the fruit, vegetables and flowers now found on the shelves of European supermarkets originate from farms in Africa. These products have become increasingly important to several African countries,[2] creating scores of new jobs for African men and women. Women, in particular, have benefited from this growth and constitute a high percentage of the workforce in African horticulture. Yet, during recent years, questions have been raised about the quality of work in global supply chains such as export horticulture that feed consumers in the North. European supermarkets are under pressure to ensure decent working conditions in their supply chains, and producers are increasingly required to comply with a number of standards and codes to guarantee that they do so. This should be good news for workers; but, in practice, what do these 'ethical trade' codes offer women workers on African farms?

In this chapter we provide some answers to this question by drawing on the findings of research carried out during 2002–2003 on farms in Kenya, South Africa and Zambia.[3] We analyse the findings in the context of a 'gender economy' framework, which recognizes that men and women experience work differently, and that their experiences are largely determined by societal norms and institutions that shape their roles and responsibilities. For instance, African women generally have to balance waged work with the expectation that they will assume primary responsibility for domestic duties. Often, their experience of work and the options available to them are directly related to the nature and extent of their domestic responsibilities. They also face discriminatory attitudes at home and in the workplace regarding the types of work they are suited to perform – attitudes that often relegate them to low-paid temporary positions. For women, therefore, the quality of horticultural work is not simply a function of the terms and conditions of employment, but is also shaped by societal perceptions and expectations of women's roles. Ethical trade operates within this context and its capacity to address the employment conditions experienced by women will depend upon the extent to which it accounts for this.

This chapter applies a gender economy approach to identify the limits and possibilities of ethical trade for women working in African horticulture. In the following section we expand upon this analytical framework. We then present the main empirical findings of the research. We conclude with an analysis of the implications of the research for gender-sensitive ethical trade.

The diversity of work within a globalized gender economy

Regulatory bodies and related institutions (such as trade unions) have traditionally formed their labour policies with a permanent, usually male, worker in mind. Such workers are increasingly scarce in a world dominated by neo-liberal economics, which advocate labour market deregulation to achieve an efficient, flexible labour force (Standing, 1999). Much work is now classified as 'temporary' or 'seasonal', and while national labour legislation may include employment benefits for these workers, these are typically fairly limited. Requirements for flexibility may also be met through the use of casual (individuals hired and paid daily on the basis of an oral contract) or contract (individuals contracted through a third party to perform specific tasks) workers. Casual and contract workers are less likely to benefit from legislated entitlements but may receive certain benefits at the discretion of employers. Importantly, all of these categories of non-permanent workers may be hired on an 'informal' basis; the terms and conditions of employment are not set out in a contract prior to starting work and workers may be denied the limited rights that they are legally entitled to. Despite the fact that in many sectors these different types of formal and informal work are becoming the norm rather than the exception, in most countries legislation has failed to keep up with these shifts in the structure of labour.

Women tend to be concentrated in non-permanent and informal forms of work, largely due to societal biases that structure the roles and responsibilities of men and women. These biases perceive men as the 'breadwinners' and women as caretakers, with women's paid work as a supplemental source of household income. The unpaid 'reproductive' work of women is a third category of work and, like informal work, has traditionally been neglected in economic analyses. Feminists have long argued that reproductive work underpins 'productive' market-based work through maintaining and reproducing the labour force, and that economic analysis should be centred around the concept of a gender

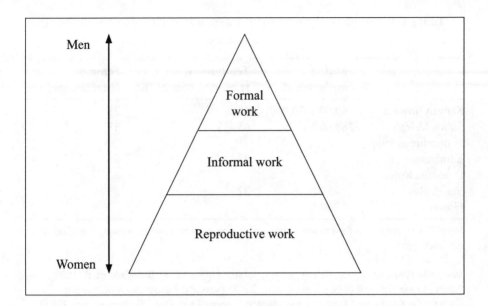

Source: adapted from Barrientos et al (2003)

Figure 5.1 *Gender pyramid of work*

economy which includes all forms of paid and unpaid work (Folbre, 1994; Elson, 1999).

The three different forms of work – formal, informal and reproductive – are depicted in the 'gender pyramid' of work shown in Figure 5.1. As illustrated in the pyramid, women are often concentrated in the lower part of the pyramid, moving flexibly between informal and reproductive work, whereas men are more likely to occupy the top part of the pyramid in permanent jobs.

Women's concentration in these forms of work directly affects the set of legislated and discretional work-related benefits that they receive and, thus, how they are likely to experience their work. Benefits vary by country, sector and company, and also by type of work, but can be divided into three categories. First, there are the formal employment rights set out in the International Labour Organization's (ILO's) core conventions and national law, including written contracts, maximum hours of work, minimum pay, freedom of association and right to collective bargaining, non-discrimination and occupational health and safety (H&S). Second, there are employment-related provisions that go beyond the core employment rights and are either legislated for, or are at the discretion of, individual employers. These include maternity and paternity leave, transport, childcare and health services. These benefits are often more important to women than men as they facilitate the combination of productive and reproductive work. Finally, there are a number of provisions that do not directly depend upon employment, but which those engaged in reproductive

Table 5.1 *Estimates of employment in export horticulture in South Africa, Kenya and Zambia*

	Total employment	Temporary, seasonal, casual (%)	Female (percentage)
Kenyan flowers	40,000–70,000	65	75
South African deciduous fruit	283,000	65–75	53
Zambian horticulture:		77	56
Vegetables	7500	32	35
Flowers	2500		

Note: The employment figures for South Africa may now be lower following a period of retrenchment.

Source: de Klerk (undated); Kritzinger and Vorster (1995, 1996); Blowfield et al (1988); Kenya Flower Council (KFC) (pers comm, 2002); National Resource Development College/Zambia Export Growers Association Training Trust (NZTT) (pers comm, 2003)

work (mainly women) may benefit from, such as health, education, childcare and community infrastructure and services. These principally fall under the remit of the state and are usually limited in availability and/or quality in developing countries. However, these provisions may also be provided by employers, particularly where workers live on site and paternalistic employment relationships prevail, as is often the case on African farms.

The relationship between the gender economy and different provisions for formal, informal and reproductive work often leaves women workers in a disadvantaged position. If ethical trade is to overcome this, it needs to be sensitive to the gender dynamics of work. But is this currently the case? In the following sections we draw on in-depth research with workers and management on 17 farms in Kenya, Zambia and South Africa to assess the extent to which ethical trade addresses the needs and concerns of women horticulture workers. These findings are set against the gendered employment context of each country, with women comprising between 35 and 75 per cent of employment in each sector/country and typically concentrated in temporary, seasonal or casual employment.

Codes of practice and women's experience of work in African horticulture

By the mid 1990s, most leading African horticulture producers were applying codes in order to satisfy the requirements of their overseas customers. These codes were introduced from four different origins: by dominant buyers such as supermarkets and importers; trade associations linked to the Northern fresh produce industry; sectoral trade associations linked to the African horticulture sector; and independent bodies comprising business and civil society organizations (Barrientos et al, 2001, 2003). While a number of different codes operate in African horticulture, not all of which strictly relate to ethical trade, almost all farms in the study were using at least one code that focused on worker welfare. In the following discussion we use the criteria contained within these codes as a way of exploring the gender-differentiated experiences of work in African horticulture.

Job security

Many ethical trade codes refer to the provision of 'regular employment' based on an established legal relationship between employer and employee. Despite fairly widespread use of contracts (albeit with limited content), the research found that job insecurity was a major concern for workers in all three countries, especially non-permanent workers. Many workers ranked insecurity as a greater concern than low wages, indicating the significance awarded to employment stability and the degree of vulnerability that workers currently experience. Although many employers said they preferred to rehire the same seasonal workers each year for continuity, workers claimed that they had no guarantee of employment from year to year. Furthermore, many non-permanent workers were on 'rolling contracts' and in reality worked year round with no guarantee of long-term employment. This practice reduces labour costs and provides employers with considerable labour flexibility; however, it is cautioned against in many codes. While insecurity affects both men and women workers, women are more likely to be found in non-permanent employment and therefore suffer disproportionately.

On the positive side, there was some evidence of an increase in permanent employment during recent years, especially in the flower sector, where women workers in particular have benefited. According to flower growers in Kenya and Zambia, employment has become more permanent for (at least) three reasons: less seasonality in production cycles; increasing need for a more stable, skilled workforce in order to maintain high quality; and pressure to implement codes. This indicates the potential role that ethical trade, in conjunction with other factors, may play in increasing the security of women workers.

Working hours

Almost all codes stipulate that regular working hours should not exceed 48 hours per week, with up to 12 hours of overtime only permitted on an occasional and voluntary basis. For women, regular, often compulsory, overtime can make it difficult to balance productive and reproductive roles. Arranging childcare when overtime is required at short notice is particularly problematic (see Box 5.1). However, the ability to work extra hours was considered to be a condition of employment in all three countries and many workers, especially women in packhouses, complained that overtime was obligatory and often excessive (e.g. up to eight hours a day). From the employers' perspective, giving prior notice of overtime is not always possible as the use of 'just in time' supply chain management means that orders must be fulfilled at extremely short notice, often within hours of the plane's departure for Europe. Compulsory overtime at short notice is therefore an important gender consideration that cannot be solved at the producer end of the supply chain alone.

Box 5.1 Overtime and childcare responsibilities

We are often told on the same day that we have to work overtime that evening. It is then our responsibility to make arrangements with the [transport] services we use; we have to pay for the phone call to change arrangements ... this is not fair – management should pay for these calls... Women who have children have to make special arrangements for their children... We are not given adequate warning to come to work prepared. (South African female packhouse worker)

Living wages

Codes typically require payment of at least the minimum or industry average wage; but some also refer to payment of a 'living wage'.[4] In all three countries, wage levels exceeded statutory minimums; yet almost all workers in the study complained that their wages were insufficient to meet their basic needs. This is reflected by the fact that although many workers experienced difficulties in working overtime, they often welcomed it as an opportunity to increase their income. Women with sole responsibility for children faced particular difficulties with low wages. For example, in Kenya, many women had migrated from rural areas to seek work, but had left their children to be cared for by their grandparents as they could not afford to keep them with them.

Yet, employers defended the wages paid by companies in the industry. Several employers pointed out that they were paying far more than other agricultural sectors and some claimed that if they were forced to pay more they would go out of business. If this is an accurate claim (and it was not within the scope of the research to investigate further), it raises questions about the sustainability of retail prices for horticulture products. It also poses wider questions on inequalities in North–South trading relationships, which are currently not addressed by codes of practice.

Employment-related benefits

Employment-related benefits such as sick pay and childcare can supplement basic wages and provide an important safety net for workers. Generally, permanent workers received a broad range of benefits, both legislated and discretionary. As a fairly recent development in all three countries, seasonal workers received many (but not all) of the same benefits on a pro-rata basis, whereas casual and contract workers received few, if any. Many non-permanent workers viewed these benefits as a major advantage of permanent work.

For women, maternity leave was particularly important; but it was rarely extended to seasonal, casual, migrant and contract workers. As a result, many non-permanent workers expressed a fear of becoming pregnant as they risked losing their jobs. Childcare facilities were another important non-wage benefit for women and their children. A significant number of mothers in Zambia and Kenya stated that older children looked after younger siblings while they worked as they could not afford childcare services or did not trust 'house girls'. This caused them concern and was an obstacle to some children's schooling. On-site childcare facilities were thus a positive feature in seven of the 17 companies (most in South Africa). However, there were some complaints from workers about the quality and hours of service provided, and in some cases facilities were only available to children living on the farm.

Box 5.2 Women, children and employment

Excerpt from notes of a focus group with women workers in Zambia:

Respondents said that some children stay home with their older siblings. One of the respondents who is breastfeeding said that she only breastfeeds her child after she knocks off. Some children do not go to school in order to look after their younger siblings. One lady complained that it is not good to leave kids alone, as no one would take care of them when they fall sick.

An important point highlighted by the research is that many employment benefits of great importance to workers are typically not covered by codes, especially sick pay, medical care and childcare. Such benefits may be included in national legislation, and codes that stipulate compliance with local laws may therefore also cover them. But the lack of direct coverage of these issues remains a weakness of many codes, particularly from a gender perspective as they facilitate women's reproductive work.

Discrimination

While all companies formally practised equal pay for equal work, gender patterns in the allocation of tasks were usually accompanied by different pay scales for predominantly 'male' or 'female' jobs. Men were more likely to be in positions considered as higher skilled, such as sprayers, irrigators and tractor drivers, which were usually permanent and better paid. Many of the 'unskilled' low-paid seasonal jobs, such as harvesting and packing, are dominated by women. Women workers were, however, valued by employers, who viewed them as better at handling fresh produce than men and often as more reliable and hard working. This perception has led some employers to revise previous policies of unequal pay, as expressed by one farm owner: 'A few years ago I paid men more than women; but when I looked at the productivity figures, I saw women outperforming men, so I equalized the wages... And they pay more attention to detail.' However, there was a general perception amongst the (mainly male) employers, and also among workers, that even with training women would not be able to supervise or manage their male colleagues.

According to most employers, promotion is based on merit and, occasionally, length of service; but several groups of workers also claimed that promotion was based on favouritism, tribalism, racial discrimination and/or corruption (money, gifts and favours – including sexual favours). Women, in particular, perceived little opportunity for training or career progression within their companies, with formal training usually restricted to irrigators, scouts, drivers, H&S officers and those handling chemicals, most of which are men. However, skills development was receiving increased attention in all three countries: in Zambia, a training institute offering professional training programmes and short courses has been established for the industry; in South Africa, several farms are developing multi-skilled teams in an effort to improve efficiency; and in Kenya, employers described training (and performance-related pay) as key to remaining competitive.

The study also found widespread discrimination against pregnant women when making decisions relating to recruitment and redundancy, and even employers acknowledged that women who are visibly pregnant are not hired. Many workers said that non-permanent women who become pregnant are asked to leave once their pregnancy advances, or are simply not given another contract when the current one ends. As a result, they experience anxiety about becoming pregnant or telling employers that they are pregnant. This form of gender discrimination contravenes most codes, yet appears to be common practice.

Box 5.3 Discrimination against pregnant women

As soon as you know you are pregnant, you know you'll be sent back home. So most of them just keep quiet about it and there have been a great number of miscarriage incidents because people are afraid to talk as they know they will lose their jobs. Permanent people are better off because they get to work until such time as they have to give birth and they come back to work, whereas temporary workers are sent straight back home and get replaced by someone else. (Female migrant worker in South Africa)

Health and safety

Codes appear to have stimulated considerable improvements in occupational health and safety, particularly with respect to the safe use of chemicals and the provision of protective clothing, toilets, washing facilities and drinking water. However, serious problems persist, with workers in all countries complaining about health problems (coughs, sore chests, skin irritation and dizziness) associated with pesticide exposure. According to workers, these problems resulted from working with freshly sprayed plants or entering greenhouses before re-entry times had expired. There were also isolated reports of spraying occurring with unprotected workers present in greenhouses. Non-permanent workers were less likely to be provided with protective clothing than permanent workers. Hand injuries from thorns were widely reported in the rose packhouses, with some piece rate workers reluctant to wear gloves since they reduce their packing speed and thus their pay. Back pain was also common amongst women workers. Pregnant women were assigned light duties in only a few companies, and workers were concerned about risks to unborn babies as a result. It should be noted, however, that two employers expressed difficulties in finding enough light duties to accommodate pregnant women, requesting guidance on this issue. Lastly, some workers complained of poor toilet and drinking water facilities, particularly in orchards and vegetable fields, and almost all said that they had no hygienic place to store their food.

Box 5.4 Women and exposure to harmful chemicals

Women are also present when spraying is being done. Some of them have children who are sucking and when they go home they just start breastfeeding their children. We don't have a place for washing our clothing or bodies after work. (Flower workers in Zambia)

Fair treatment

Ethical trade codes often explicitly prohibit the use of harsh or inhumane treatment in the workplace, including sexual harassment. The research found that relationships between workers and supervisors were often poor, with complaints of verbal and occasional physical abuse, dismissal without just cause, wages docked as a disciplinary measure, corruption and favouritism. Non-permanent workers were particularly vulnerable to this abuse, as supervisors were often responsible for either hiring them or influencing whether they were hired, and they feared not being rehired if they complained or resisted. Like other aspects of their employment, workers were often unclear about company rules and regulations regarding disciplinary procedures. Most workers, especially non-permanent workers, felt that they could be fired without notice or just cause, and had little opportunity to appeal the decision. With high unemployment in all three countries and few alternative sources of income, the priority of workers was to keep whatever job they had.

Box 5.5 Harsh treatment in the workplace

We are only allowed to go during [designated] toilet breaks or lunch and tea breaks. Many of the workers complain of bladder infection. Once, one of the workers, who was pregnant at the time, wet herself because the supervisor did not give permission for her to go to the toilet. She had to go home and change, and what's even worse is that she wasn't paid for the hour that she took to go home. (South African women packhouse workers)

In 11 of the 17 companies in the study, workers reported sexual harassment, for the most part referring to male supervisors who demanded sexual favours in exchange for promotion, pay rises or continued employment. Women's concentration in non-permanent work makes them especially vulnerable to such abuse since the renewal of their contracts often depends upon supervisor recommendations. While most employers were aware that sexual harassment could be a problem, the majority didn't believe that it occurred in their own companies.

Box 5.6 Sexual harassment

When a male supervisor seduces a female worker and this doesn't bear fruit, he can use 'thorax' [i.e. job power] to win that female worker. (Kenyan male worker)

Freedom of association and the right to collective bargaining

According to the core ILO conventions, workers have the right to join a trade union or any other association of their choice and to bargain collectively. These rights are usually reflected in ethical trading codes. Although half of the companies in the study were unionized, less than 50 per cent of permanent workers were members, and non-permanent workers were only unionized in two companies (in Zambia and South Africa). Many casual, seasonal and migrant workers had little knowledge of unions and were unclear about their potential benefits. Trade union officials said that the low wage levels in agriculture made recruitment difficult, partly because workers were reluctant to pay membership fees and partly because fees had to be kept at such a low level that resources for recruitment drives (and training) were limited. This was confirmed by workers, with some claiming that unions either did not, or could not, bring about significant improvements. Some workers believed that management was opposed to unions and that membership could lead to loss of employment or to management intimidating shop stewards to such an extent that they were rendered ineffective. Interviews with management did reveal some negative attitudes in relation to unions, and there was little indication that codes were effective in changing attitudes such as these.

The majority of farms had workers' committees; but many workers were either unaware of their existence or felt that they were ineffective, especially in South Africa and Zambia. This was particularly true for non-permanent workers, who were generally not represented on workers' committees. However, in Kenya, one-third of workers found that committees had been influential in improving working conditions in a number of ways.

Women were under-represented in unions and workers' committees despite the fact that women often constituted the majority of the workforce. This creates a vicious circle, with the workers most in need of union representation (i.e. low-waged, non-permanent and women workers) often the least likely to be members. Both men and women workers in Kenya expressed the opinion that there were more men than women elected to committees because men made better leaders. For example, one woman said: 'Women are scared of asking questions and therefore [are] not good representatives' and another stated: 'Men champion better the workers' rights as they do not fear questioning.' This lack of confidence in women's abilities demonstrates the importance of providing support for women to voice their opinions. In recognition of the fact that women lacked a forum to express their views, one company in Kenya had established a gender committee, and two others (in Kenya and South Africa) stipulated that workers' committees should be made up of equal numbers of men and women.

Grievance mechanisms

Senior management was clearly unaware of many of the problems that employees faced. The hierarchical organizational structure of most companies was quite rigid, relying on line management and vertical communication channels. As such, general workers had few, if any, opportunities to interact with senior management and relied on their supervisors to pass information both up and down the line. Supervisors often created a block in this communication channel, with workers left uninformed about company policies and management unaware of workers' grievances. This was complicated by the fact that many non-permanent workers were fearful of complaining, believing that if they did so they would be viewed as troublemakers and not rehired. Language was also an obstacle to communication, with some workers in all countries complaining about written and/or verbal information provided to them in a language they did not understand well. In general, workers felt that they did not have a secure and effective channel through which they could communicate their grievances and achieve redress.

There is a significant gender aspect to this situation as the complaints or queries of women are usually required to pass through a series of predominantly male managers before reaching senior managers or company directors. This decreases the likelihood that women workers will report sensitive issues and increases the chance that their comments will be lost or distorted.

African horticulture, the gender economy and ethical trade

The perspectives of women workers in African horticulture, as outlined above, tell a clear story about the way in which their experience of work is embedded in the culture and norms of their societies, which assign particular roles and responsibilities on the basis of gender. Women face a number of difficulties arising from their need to combine productive with reproductive responsibilities. These include struggling to pay for childcare out of a low wage packet, and to arrange childcare at short notice when they are required to do overtime, as well as insecurity and health concerns due to discrimination against pregnant women. However, where the gender disparity becomes particularly pertinent is in relation to women's engagement in more short-term, less formal types of waged work. Such workers are less likely to have access to company childcare services and protective clothing, and are more likely to suffer from harsh and inhumane treatment, including sexual harassment. They are also rarely represented in workers' organizations, have few channels for communicating their grievances, and are reluctant to lodge complaints for fear of not being rehired. To compound matters, they are sometimes offered little protection by national law.

Referring back to the different work-related provisions we described earlier, we can see that formal rights mainly accrue to workers hired on permanent contracts, although temporary and seasonal workers on formal contracts also benefit, depending upon national legislation (see Figure 5.2). There are also some spill-over effects on informal workers. For example, in some cases hours of work and wages are standardized across the labour force. The receipt of employment-related benefits such as childcare provision, maternity leave, transport and housing often depends upon company policy, and short-term and informal workers are excluded in many instances. As noted earlier, these benefits are particularly important to women, who are concentrated in informal employment and who are primarily responsible for combining paid work with reproductive labour. In the case of benefits related to reproductive work, the beneficiaries are all of those living in a community, regardless of employment status; but it is often the case that only permanent workers are provided with on-site housing.

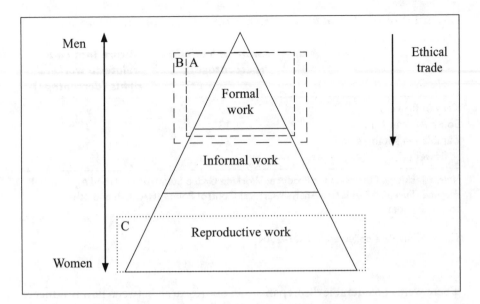

Notes: A = Regulation and provision of formal conditions of employment.
     B = Regulation and provision of employment-related benefits.
     C = Non-employment-related benefits and social provision supporting
        reproductive work.
Source: adapted from Barrientos et al (2003)

Figure 5.2 *Gender pyramid, work-related benefits and ethical trade*

Can codes deliver benefits to women workers?

Most producers in this study recognized the importance of good employment practices and seemed genuinely committed to making improvements. However, they often had a limited understanding of the problems that workers faced. While codes have raised awareness of certain issues, such as health and safety and working hours, in general, producers were instituting changes based on their own perceptions and not those of workers. Very few workers in our study had any knowledge of codes, despite the fact that some companies had been implementing them for several years. Those workers who were aware of codes generally thought that they related to technical specifications, with only a few aware that codes covered workers' rights (see Table 5.2).[5]

Furthermore, while codes have contributed to some improvements in African horticulture, particularly in the area of health and safety, many of the gender issues outlined in this chapter have persisted. This is partly due to the prevalence of 'checklist' and 'snapshot' auditing, which has overlooked the concerns of marginalized workers and failed to identify sensitive issues such as gender discrimination and sexual harassment (Barrientos et al, 2003). This is particularly the case where the employer is a principal source of information.

Table 5.2 *Percentage of workers aware of codes and what codes are*

	Aware of 'codes' (percentage)*	Aware that codes relate to workers' rights (percentage)
Kenyan flowers	22	7
South African fruit	13	3
Zambian vegetables and flowers	31	4

Note: * Including Euro Retailer Producer Working Group Standard for Good Agricultural Practice (EurepGAP), Hazard Analysis Critical Control Points (HACCP) and other technical codes.

Source: Study data collected in 2003–2004

There were considerable discrepancies in our research between how producers perceived employment conditions and what their employees reported. Even if auditing/monitoring were not seen by producers as a threat, there is a high probability that the information they provide will not paint an accurate picture of the reality for workers themselves, particularly in relation to gender issues. Some non-permanent workers in our study, who did not have proper protective clothing, even reported that when visitors (e.g. buyers and auditors) came they were told to hide.

If codes are to enhance the rights of all workers, including women, they must be implemented in a gender-sensitive way. Participatory social auditing, where workers are integrated within every stage of the auditing process, including (and especially) during feedback and decisions regarding remediation (see Chapter 8), is seen as one way of improving the effectiveness of codes. Our study found that participatory social auditing methods are not only effective in discovering areas of non-compliance that are not detected by other methods, but are especially helpful in identifying the needs of non-permanent workers, who are typically excluded in formal auditing approaches and often face the poorest working conditions.

Yet, participatory social auditing can only be one strategy in a more comprehensive process of workplace improvement. Ensuring workers' rights requires leveraging a range of mechanisms, including national legislation, international labour standards and the efforts of various stakeholders. There is growing recognition, including among several European buyers, that a multi-stakeholder approach to code implementation, incorporating *local* private sector, civil society and government bodies, may have greater potential to effect workplace improvement, particularly in ensuring the representation of marginal workers and women. The Wine Industry Ethical Trade Association (WIETA) in South Africa and the Agricultural Ethics Assurance Association of Zimbabwe (AEAAZ) are examples of locally owned multi-stakeholder initiatives that have led to a notable change in working conditions (see Chapter 8). These initiatives can help to offset the limitations of code content and auditing, and can also provide African governments with an important vehicle through which to support national and international labour legislation.

At the same time, Northern stakeholders must also play a role in improving workplace conditions in global supply chains such as export horticulture. For retailers, such efforts could include reforming their purchasing practices so that producers no longer deflect the pressures of low prices and 'just-in-time' delivery onto their workforce. Buyers could also include coverage of non-permanent workers in company codes and assist their suppliers to meet the costs of code compliance. In particular, support for participatory social auditing and local multi-stakeholder initiatives, as well as the provision of guidelines and education on gender issues, could potentially transform what has, to date, been a largely gender-neutral approach into one that incorporates the gendered needs and rights of all workers.

Conclusions

Pressures for ethical trade have forced global buyers to apply codes of practice that address the working conditions in their supply chains. These codes are meant to ensure that all workers are provided with safe and fair working conditions. Yet, codes have been introduced into a social and economic context with deeply embedded gender inequalities. As this chapter has shown, the environment in which codes operate is not gender neutral, but is shaped by the gender economy, predicated on a division of labour between productive and reproductive work. Women's primary responsibility for reproductive work constrains their access to formal paid employment, increasing their concentration within informal work. It also conditions their specific gender needs as workers as they must juggle their reproductive responsibilities in the context of the insecurity created by informal work. They face intensified gender risks, such as lack of maternity leave and childcare, and do not have adequate job security or employment protection to cope with those risks.

A gender economy approach reveals the limitations that codes face in enhancing the working conditions of women and other marginalized workers. As noted, while codes address the employment needs of formal workers, they often fail to tackle the more complex gender needs of informal workers, where the conditions of employment are often worst. If codes are to address the employment issues faced by the majority of informal workers, their gender sensitivity will need to be enhanced. This can be achieved in a combination of ways. First, the content of codes can be extended to include employment-related benefits, such as childcare, maternity leave and transport, which reduce the stress and insecurities women experience in combining their reproductive roles with paid work. Second, the scope of codes could be extended to include non-permanent workers within their remit, broadening the access of all workers to the employment protections found in codes. Third, the auditing process could be used as a way of identifying workers' needs and concerns, and of integrating workers themselves directly within the process of workplace improvement.

Yet, irrespective of how the process of code implementation is reformed to address the needs of workers, it is imperative that all actors in the supply chain, including government, civil society and the private sector, assume responsibility for ensuring labour rights in global industries. Ultimately, it is only by cultivating the engagement of various stakeholders and drawing on all forms of employment regulation that the conditions faced by women workers in African horticulture can be ameliorated.

Notes

1   This chapter is based on a jointly authored paper written in association with D. Auret, S. Barrientos, K. Kleinbooi, C. Njobvu, M. Opondo and A. Tallontire (see Smith et al, 2004).
2   In Kenya, for example, horticulture is the fastest-growing sector of the economy, generating over US$270 million and accounting for 22 per cent of all agricultural exports in 2000 (Gachanga, 2002), while in South Africa the total value of deciduous fruit exports alone stood at US$700 million that year (Deciduous Fruit Producers Trust, pers comm, 2000). In the case of Zambia, one of the poorest countries in Africa, year-on-year growth of horticultural agricultural exports exceeded 40 per cent, on average, during the late 1990s (Giovanucci et al, 2001).
3   This paper is based on a research project funded by the UK Department for International Development (DFID) (SSR Project 8077, Ethical Trade in African Horticulture), coordinated by the Institute for Development Studies (IDS). DFID supports policies, programmes and projects to promote international development. It provided funds for this study as part of that objective; but the views and opinions expressed are those of the authors alone.
4   A living wage is remuneration sufficient to meet workers' basic needs, such as food, housing, transport, healthcare, education and clothing, as well as to provide some discretionary income (see the Ethical Trading Initiative, www.ethicaltrade. org/Z/lib/2000/06/livwage/index.shtml#liv-fair).
5   Lack of worker awareness of codes does not necessarily mean that codes have no impact; but it constrains deeper, more long-term improvements in workers' rights.

References

Barrientos, S., Dolan, C. and Tallontire, A. (2001) *Gender and Ethical Trade: A Mapping of the Issues in African Horticulture*, Natural Resources Institute, Chatham Maritime
Barrientos, S., Dolan, C. and Tallontire, A. (2003) 'A gendered value chain approach to codes of conduct in African horticulture', *World Development*, vol 31, no 9, pp1511–1526
Blowfield, M., Malins, A. and Dolan, C. (1988) *Kenya Flower Council Support to Enhancement of Social and Environmental Practices: Report of The Design Mission*, Mimeo, Natural Resources Institute, Chatham Maritime

de Klerk, M. (undated) *Deciduous Fruit Industry Study*, Commission of Inquiry into the Provision of Rural Financial Services, Cape Town, South Africa

Elson, D. (1999) 'Labor markets as gendered institutions: Equality, efficiency and empowerment issues', *World Development*, vol 27, no 3, pp611–627

Folbre, N. (1994) *Who Pays for the Kids? Gender and the Structures of Constraint*, Routledge, London

Gachanga, S. P. (2002) 'Horticultural industry in Kenya', Paper prepared for the Globalization, Production and Poverty Workshop, Nairobi, Kenya

Giovanucci, D. P., Sterns, P. A., Eustrom, M. and Haantuba, H. (2001) *The Impact of Improved Grades and Standards for Agricultural Products in Zambia*, Phase One Assessment and Recommendations for USAID, Institute for Food and Agricultural Standards, Lansing, Minnesota

Kritzinger, A. and Vorster, J. (1995) *The Labor Situation in the South African Deciduous Fruit Industry*, Research Report, University of Stellenbosch, Stellenbosch, South Africa

Kritzinger, A. and Vorster, J. (1996) 'Women farm workers on South African deciduous fruit farms: Gender relations and the structuring of work', *Journal of Rural Studies*, vol 12, no 4, pp339–351

Smith, S., Auret, D., Barrientos, S., Dolan, C., Kleinbooi, K., Njobvu, C., Opondo, M. and Tallontire, A. (2004) 'Ethical trade in African horticulture: Gender, rights and participation', *IDS Working Paper 223*, Institute of Development Studies, Brighton

Standing, G. (1999) 'Global feminization through flexible labor: A theme revisited', *World Development*, vol 27, no 3, pp583–602

6

Central American Banana Production: Women Workers and Chiquita's Ethical Sourcing from Plantations[1]

Marina Prieto-Carrón

Introduction

Plantation agriculture has long had a reputation for poor labour practices. Yet, over the last decade, several multinational corporations in the banana industry have come under increasing pressure by consumers and buyers to improve labour conditions, and most have initiated various programmes in the area of corporate social responsibility (CSR). As part of this process, many corporations now apply workplace codes of conduct as voluntary policy tools to regulate labour conditions in their workplaces.

This chapter focuses on Chiquita, one of the world's largest producers and marketers of bananas, who was rated as one of 1995's ten worst corporations by *Multinational Monitor* (Mokhiber and Wheat, 1995). Over the last decade, Chiquita has taken a number of steps to rectify this image and improve its ethical performance through the development of a comprehensive CSR policy. One important aspect of Chiquita's ethical sourcing is its voluntary code of conduct and monitoring, which company-owned plantations and independent producers must implement. This chapter examines the challenges and opportunities that Chiquita faces in implementing its code and its overall CSR policy. In particular, it highlights women's experiences of working for Chiquita (on both owned and subcontracted plantations), and describes how Chiquita has responded to the conditions that they face through its CSR policies. The chapter argues that the capacity of CSR initiatives to benefit women workers is not only related to the corporate policies of Chiquita, but to the structural conditions of the international political economy, which produce different experiences of work among men and women.

The material presented in the chapter is based on two focus group discussions conducted in July 2001 with women banana workers in the local trade union, Associación de Trabajadores del Campo (ATC), in Chinandega, Nicaragua; interviews with representatives of the trade union; and an in-depth interview conducted with George Jaksch, Chiquita's senior director of corporate responsibility and public affairs.[2] This material is supplemented with the questionnaire responses of two anonymous Chiquita managers from Costa Rica in 2004, and information drawn from secondary sources, including material from Chiquita and various organizations.

The next section provides a brief introduction to Chiquita's ethical policy and code of conduct, and is followed by background on Nicaragua's position in the international banana industry. The chapter then focuses on the issues faced by women banana workers in Nicaragua and Chiquita's code of conduct. The last section of the chapter describes how Chiquita has responded to the concerns of women workers and the current challenges that the company faces to improve workers' rights in the banana industry.

Chiquita's ethical policy and code of conduct

Chiquita has introduced a number of CSR initiatives in its production and sourcing operations. In 1998, the company established a steering committee and a senior management group to steward CSR activities, and in 1999 it adopted core values. During 2000, it adopted a code of conduct, began management training, appointed a corporate responsibility (CR) officer reporting to the president and board, began internal assessments, achieved a 100 per cent Better Banana Project Certification in its own divisions (Guatemala, Honduras, Costa Rica, Panama and Colombia), and claims to have influenced 30 per cent of the independent producers to adopt the standard (Zaffa, 2001). In 2001, Chiquita instituted worker training and 'broke ranks with the other multinationals', signing an agreement with the Latin American Banana Workers' Union Coordination (COLSIBA) and the International Union of Food, Agricultural, Hotel, Restaurant, Catering, Tobacco and Allied Workers' Associations (IUF) on 14 June 2001, committing the company to respect core International Labour Organization (ILO) conventions (Chiquita et al, 2001). By the end of 2002, Chiquita estimated that 87 per cent of its farm workers were union members, the highest proportion among major banana growers (Chiquita, 2003). During the same year, Chiquita joined the multi-stakeholder UK-based Ethical Trading Initiative (ETI), and by 2005, it had achieved Social Accountability (SA) 8000[3] standard certification on its owned farms in Latin America, which provide employment to 14,000 men and women.

Chiquita's code of conduct, which applies to its own plantations and to suppliers, reflects its goal to conduct business 'in an ethical, legal and socially responsible manner befitting a world-class company' (Chiquita, 2000, p1). According to Chiquita: 'Our final aim is to give all our business to those suppliers that comply with the social responsibility in our code and that behave in an ethical and legal manner' (Chiquita, 2000). The code covers the main labour issues in national law and relevant international conventions, and sets forth four core values – integrity, respect, opportunity and responsibility – for issues ranging from compliance with labour laws to insider trading and political donations. The section covering workplace practices incorporates the SA 8000 standard almost verbatim, with a few additions. Thus, the company is committed to respecting a range of international principles embodied in the Universal Declaration of Human Rights, the United Nations Convention on the Rights of the Child, and various ILO conventions. It is also committed to complying with national legislation, stating that:

> ...the company shall comply with national and applicable law, other requirements to which the company subscribes, and this standard. When national and other applicable law, other requirements to which the company subscribes, and this standard address the same issue, that provision which is most stringent applies.
> (Chiquita, 2000, p7)

While there are some weaknesses and ambiguities in code content, Chiquita's monitoring practices have been well received. The company produces detailed reports outlining areas of compliance and non-compliance, thus presenting an image of balanced reporting. Similarly, the company bulletin, *Corporate Responsibility News*, provides specific information on CSR activities and outlines the progress that the company has achieved in fulfilling its objectives. These efforts have garnered praise for Chiquita in several quarters. The magazine *Ethical Performance*, for example, identified Chiquita as an example of best practice in 2001, and Juan Somavía, director general of the ILO, witnessing the signing of the labour rights agreement with the IUF and COLSIBA, noted that 'new ideas always need pioneers, and you are the pioneers of something that I believe is going to be run in the future' (Zaffa, 2001).

Yet, Chiquita also recognizes that social problems exist and has acknowledged so publicly in corporate reports and at various events in which the company participates. For example, a company representative described his surprise upon discovering divisions that did not allow breastfeeding breaks and that performed pre-employment pregnancy and AIDS testing (Zaffa, 2001). There have also been several accusations of union rights infractions. In September 2003, Chiquita subsidiary Compañia Bananera Independiente de Guatemala SA (COBIGUA – Independent Banana Corporation of Guatemala) closed four unionized farms in Guatemala, citing low productivity and oversupply. Yet the trade union, Unión Sindical de Trabajadores de Guatemala (UNSITRAGUA – The Guatemalan Workers' Union) maintains that the real reason for plant closure was the transfer of production to non-union plantations and charges that there was a failure to consult with trade unions (as the framework agreement specifies) (Banana Link, 2003b). Hence, despite its visible efforts and espoused commitment to fair working conditions, the company continues to face difficulties instituting its CSR policies (including the framework agreement with COLSIBA and IUF), and ensuring compliance (see, for example, the ILO study of the framework agreement in Riisgaard, 2004). As the next section illustrates, however, these difficulties are compounded by the structural conditions in the international political economy.

The political economy of the banana industry

The Nicaraguan banana industry

The export of bananas from Nicaragua dates from 1950, when bananas were under the management of the US-based United Fruit Company (which subsequently became Chiquita). United Fruit managed production of the fruit until the Sandinista Revolution (1979–1990), when the state took an active role in administering banana properties. In 1981, the government of Nicaragua signed an agreement with Dole, enabling the company to take control of technical assistance and quality control of production and export (mostly to the US). In 1982, due to political pressure and a US trade embargo, Dole withdrew from Nicaragua and the nationally owned Nicaraguan Banana Producers' Association (BANANIC) assumed control over production and trade with the aim of penetrating the European market.

In 1993, with the Sandinistas no longer in power, the US lifted the embargo and Dole returned to Nicaragua. During this period the country underwent wide-spread privatization, including the banana sector. An agreement was reached between the government, banana workers and companies whereby the workers owned 25 per cent of the BANANIC export licences, while the remaining 75 per cent were held by private producers. The workers established their own company, the Nicaraguan Banana Workers' Association (TRABANIC), with 4352 partners. The structure of the industry was also notably different from the pre-revolution period, with locally owned companies exclusively supplying a transnational company.

In 1999, Dole terminated its agreement with Nicaragua and Chiquita began purchasing from local private producers, now termed 'independent producers'.[4] This type of subcontracting, where transnationals use an intermediary company to employ workers on their behalf, has been a feature of banana production to greater or lesser degrees throughout Central America. According to the local trade union ATC in Chinandega, Nicaragua, the subcontracting system allows transnationals to skirt the payment of social security and other benefits, a practice well established in the horticultural industries in other parts of the world (see Chapters 5 and 7).

Since the early 1990s, the banana industry has been hard hit by changes in the global economy. However, by the end of the decade a number of factors converged to create serious challenges for Nicaraguan production and trade. Between 1998 and 2002, for example, Nicaragua's banana trade was severely affected by climatic problems and the devastation wrought by Hurricane Mitch, as well as by political disputes, overproduction and falling prices. To compound matters, in 2002, Dole and Chiquita became locked in a legal dispute over Dole's ownership of licences for European imports, which was effectively barring Nicaraguan exports to Europe. TRABANIC and BANANIC charged that they were denied use of their licences and that the lives of many workers and their families were being destroyed. At the same time, local trade unionists were publicly denouncing the social disintegration and downward pressure on wages associated with the closure of banana plantations.

The international banana trade

The conditions faced by Nicaraguan banana workers, in general, are among the worst in Latin America (Riisgaard, 2004).[5] The confederation of Latin American banana unions (COLSIBA) has reported that years of legal wrangling – for example, at the World Trade Organization (WTO) – coupled with overproduction, have negatively affected the pay, working conditions and security of banana workers, and contributed to mass unemployment in traditional banana-producing regions. These conditions, however, must be seen within the wider socio-economic context in which Structural Adjustment Policies (SAPs), privatization and liberalization have created a number of negative consequences for the country, including unemployment, low wages, labour rights violations, retrenchment and flexibility.

The oligopolistic structure of the industry has exacerbated these conditions. A handful of vertically integrated transnational corporations (Chiquita, Dole, Del Monte Fresh Produce, Noboa and Fyffes) currently dominate international banana marketing and trade (Vorley, 2003). In 1999, three companies (Chiquita, Dole and Del Monte) controlled 65 per cent of the market, with market shares of 25, 25 and 15 per cent, respectively, illustrating the scale of benefits captured by developing countries (UNCTAD, undated).[6] This concentration is particularly apparent in the structure of the Latin American/Caribbean banana trade to the UK, where there are 60 million consumers; 5 retailers with 70 per cent of the UK grocery market; 5 companies or alliances of ripeners or distributors with 88 per cent of the UK market (includes Keelings/Chiquita, Fyffes, Del Monte, JP/Dole and S. H. Pratts); 5 transnational banana companies that control 80 per cent of the world market; and 2500 plantations, 15,000 small- to medium-scale farmers and 400,000 plantation workers in the export sector (Vorley, 2003, p51).[7]

Yet, despite their market share, banana multinationals are under increasing pressure from supermarkets and retail chains in the US and Europe. These supermarkets have consolidated and assumed greater control over production and distribution. They have also established long-term relationships with preferred suppliers and thus are able to reap higher profits without taking direct ownership over production. In the UK, Wal-Mart-owned Asda has provided an exclusive contract to Del Monte, lowering the price from UK£1.08 to UK£0.85 per kilo in 2003. In order to remain competitive, other supermarkets have been forced to follow suit even if they have to sell bananas at a loss (Banana Link, 2002, 2003a; Ryle, 2003).[8]

The buying policies of these retailers have forced suppliers such as Chiquita to seek ever-cheaper sources of supply, creating competition between Latin American countries to offer lower prices (see also Raworth, 2004). Yet, the costs of banana production are contingent upon the minimum prices set by governments, as well as upon prevailing labour costs. For example, the prices of bananas from the Windward Islands are generally higher due to higher labour costs and better labour conditions. By contrast, Ecuador, Latin America's largest producer, is able to supply bananas at comparatively low prices due to overproduction and poor labour conditions. Multinationals argue that they can only survive with low-cost production and that high costs in some countries create market distortions.[9] Yet, for workers and producers, who face periods of unemployment and/or a worsening of their labour conditions, market saturation and low consumer prices have exacerbated their vulnerability.

Gendered political economy: Women workers and labour conditions

In order to maintain their competitiveness in an international political economy characterized by increasing liberalization and corporate consolidation, companies often depress wages and working conditions. Yet, this is not a gender neutral process; as studies of the global fresh produce industry show, it is often women who suffer most (see Barrientos et al, 1999; Dolan, et al, 2003; Smith et al, 2004). This section explores the conditions faced by women working in the banana industry and describes their perspectives on the Chiquita code of conduct. As Iris Munguia, coordinator of the Women's Secretariat of COLSIBA and one of the best-known advocates of women workers' rights in the banana sector said in an interview with the *Big Issue*, 'how much difference it made just talking to women as if they really mattered' (Harris, 2003, p32).

There are approximately 482,000 women working in the banana industry in seven Latin American countries (Guatemala, Honduras, Nicaragua, Costa Rica, Panama, Colombia and Ecuador) (COLSIBA, 2001, p13). The majority of these women are involved in packing, a job for which women are seen as ideally suited, and face a similar set of working conditions (Aseprola, undated; STITCH, undated; Naranjo, 1999; Bendell, 2001;[10] COLSIBA, 2001). In general, the conditions faced by women across the banana industry are more insecure and taxing than those faced by men. Women often work an average of 14 to 16 hours a day, many while being the sole caretaker for families of five members or more, earning low wages. Typical daily wages in banana packing houses are US$1.20–$1.50 in Nicaragua, US$4.50–$6 in Guatemala and US$7–$10 in Costa Rica (Banana Link, 2004, p4). Women also face a number of occupational health issues stemming from long hours in the packing houses, including skin allergies and cancer from pesticide exposure, and injuries caused by long hours of fast-paced repetitive labour. These are compounded by the psychological effects of excessive control by management (including threats against organization), isolation from family, poor or non-existent maternity benefits and pervasive sexual harassment. According to Iris Mungia, sexual harassment is justified by some producers as 'part of the culture' (Banana Link, 2004, pp3–5), but it also is an instrument of control used widely on banana plantations (Enloe, 2000, p140). In Nicaragua, for example, sexual harassment is commonly practiced during workers' 60-day trial period, where women are often pressured to sleep with supervisors in order to retain their jobs.

Work on banana plantations is also highly insecure as the seasonal nature of the industry means that for many women work is only available when fruit is harvested. Even during the harvest period, however, many women do not have stable work hours; rather, they work 'on call' depending upon the work available. This means that women, who assume the majority of reproductive responsibilities, are obliged to take on additional income-generating activities such as domestic service and the casual selling of food and other items. The banana industry also provides women with few skills to transfer to new employment when they are forced to leave the industry at the age of 35 to 40. In addition, in many areas, the packing house offers the only prospect of employment; thus, women are forced to migrate, with all the personal and social costs that this implies.

These issues are often exacerbated by the wider social context in which domestic violence is widespread and women are negatively perceived in the prevailing culture (COLSIBA, 2001, p36). These countries have a history and culture of gender relations that, despite differences among women, sustain gender stereotypes experienced by many women all over the world: the gendered segregation of work; women as second earners and men as breadwinners; the unequal pay for equal work; women as solely responsible for the household and children; and the invisibility of women's work (ILO, 2004).

Nicaraguan women workers

In the focus groups conducted in 2001, Nicaraguan women workers reported issues similar to those documented in other export horticulture industries, with some local specificities (see Table 6.1) (Smith et al, 2004; see also Chapters 5 and 8). With the lowest wages in the region, pay rates were a major grievance, along with unpaid benefits, overtime and docking of wages for disciplinary reasons. This low remuneration is linked to short-term contracts, very long hours and difficulties obtaining sick leave, all of which make it hard to balance productive and reproductive roles. Health and safety issues also occupied a large part of the discussion, especially the consequences of agrochemicals on reproductive health, such as sterility, malformed children and various cancers. Workers complained that the water used for cooking, washing and subsistence farming was contaminated and created specific problems for women workers when pregnant or breastfeeding. Women also reported obstacles to freedom of association and harassment from trade unionists on banana plantations.[11]

The findings of the focus group discussions suggest the need for gender-sensitive practices to improve the conditions of women workers. Sexual harassment, for example, could be addressed by measures such as employing

Table 6.1 *Women workers' issues*

Women's issues	Women workers' testimonies
Discrimination	Clear sexual division of labour, with no women working in supervisory roles Discrimination in hiring, with women having to take pregnancy tests to gain employment Few opportunities for upward career progression
Verbal and sexual harassment	Verbal harassment widely reported with psychological pressure, often connected to pressures for high productivity Sexual harassment endemic, linked to cultural institution of machismo Harassment is exacerbated by the prevalence of temporary/insecure work
Childcare and maternity rights	Many women are single mothers Older children care for younger siblings, curtailing their own education and opportunities, or domestic helpers use a big proportion of already insufficient earnings Maternity rights are violated (maternity benefits, breastfeeding, etc.)
Safe transportation	Women feel vulnerable when they leave their shifts at night as transportation is not provided and they must often walk long distances

Source: Author fieldwork

more women in supervisory and management roles, creating opportunities/ spaces where women can safely report incidents, and adopting a zero tolerance policy towards perpetrators. Another key issue is raising gender awareness by training male workers and supervisors, and developing policies to combat societal norms that contribute to ongoing sexual harassment. Companies could also work creatively with women and men workers to find ways of providing convenient and affordable childcare facilities. Ensuring that both female and male staff have access to appropriate information is the first step in guaranteeing that maternity rights are respected; that female staff are not questioned about or tested for pregnancy (either before or during employment); that adequate provisions are made for pregnant workers to remain healthy and safe at work and to receive their legal benefits; and that female workers are not dismissed if they become pregnant.

Yet, a change in employment culture will necessitate the engagement of different actors at local, regional and international levels. An important actor is Chiquita. In the following section, women workers speak about Chiquita's code of conduct and how they might leverage Chiquita's ethical policy to work for them. As these discussions highlight, women are not simply passive victims, but are willing to struggle for their rights.

Women workers and Chiquita's code of conduct and monitoring

Chiquita's code states that its own employees should be aware of the code of conduct. However, given the nature of subcontracted work on plantations, this can be difficult. Only a few of the women in the focus groups had heard of the concept of a corporate code of conduct (largely those who were active in the trade union), and none of them knew what the Chiquita or SA 8000 code contained. As one worker noted: 'If you had not come with this code, we would have continued without knowing of it.' When the code was first explained to the women, there was some disquiet. They were amazed at how different their working conditions were from those described in the code, and were somewhat sceptical of Chiquita's intentions. However, after further discussion some workers began to consider how the code might help to improve their situation and how they themselves might use the code to further their interests. As one worker said: 'Chiquita's code could improve our situation', while another suggested that 'knowing the code is going to remind us of the necessity of being organized in order to begin to claim our rights'.

The women also suggested that the most important step would be to raise awareness of Chiquita's code (as well as other codes) among workers and their organizations. As one worker claimed: 'We don't know our rights.' They suggested that training be provided on the code and on aspects of particular concern to them, such as health and safety, as well as assertiveness training to confront discrimination and harassment.

Women felt that it was necessary to put pressure on managers to acknowledge the code and their obligations to comply with it. One woman said that when she raised the issue of Chiquita's code, the plantation manager claimed that they were not working with Chiquita despite the fact that all Nicaraguan-exported bananas were supplied to Chiquita at the time. In general, there was a widespread perception that Chiquita could do more to influence the behaviour of its suppliers. As one woman said: 'Chiquita has influence in order to demand quality from the suppliers, but they also have influence so the suppliers can have quality towards the worker.' One ATC leader also argued that 'They [Chiquita] say that they are not producers here in Nicaragua; but they are buyers and they give technical assistance, so they also can put forward a more just treatment of workers.' Women felt that Chiquita and the management of the independent plantations should engage with workers to define what the code meant in specific circumstances. A number of women suggested that the local unions should take the lead in these discussions, with a permanent joint commission established between Chiquita and the unions to interpret the code and agree on monitoring and implementation procedures.

The issue of monitoring was particularly important to workers. While Chiquita's code of conduct embraces a commitment to monitoring, women had little to no direct engagement with monitors. Women described their experience of government labour inspectors as poor, recounting stories of inspectors coming to have lunch with management and leaving without speaking to the workers, or foreigners who 'just take our photos while we work'. They also recounted quality audits performed by Chiquita corporate staff where everything is carefully organized prior to the inspection; but 'the following day after Chiquita's visit we return to the regime of [before]'. Given these experiences, workers recommended that monitoring of the code take place without prior warning or coordination with plantation managers and that monitors speak directly to workers. They also believed that monitors should approach the workers off site, perhaps via the trade union.

Chiquita's ethical policy responds to women workers

These focus group discussions with women workers highlight the weaknesses in implementing the Chiquita code of conduct among independent producers, particularly with regard to gender issues. These weaknesses relate to the lack of awareness of codes among workers, the minimal commitment of local managers and the superficial monitoring techniques currently practised. However, Chiquita's ethical performance is better on its company plantations, where its own divisions have instituted a number of positive actions towards improving working conditions. For example, Chiquita is working with monitoring groups in the region (Commission for the Verification of Codes of Conduct – COVERCO – and Salvadorian Independent Monitoring Group – GMIES) to provide external observers in audits, including women sensitive to gender issues, and has opened various spaces for local trade unions. According to managers of Chiquita in Costa Rica, all workers are able to report non-conformance against the code of conduct or SA 8000 in mailboxes located in all work centres. Chiquita has also widely distributed a popular version of the code to train workers (17,000 workers by the end of 2002) and has taken actions to address violations of labour rights as described in their 2000, 2001 and 2002 corporate responsibility reports.[12] The following section examines Chiquita's response to some of the gender concerns Nicaraguan women workers raised and explores the challenges Chiquita faces in achieving long-term improvements in women workers' lives.

Chiquita on women workers' specific issues

Company codes do not address a number of issues that are important to women. To some extent this reflects the fact that codes are based on a liberal feminist perspective[13] that aims to ensure equal treatment of men and women (and this is also a challenge). Nevertheless, the Chiquita code covers a number of issues that are significant to women. It prohibits all forms of discrimination – 'the company shall not engage in or support discrimination in hiring, compensation, access to training, promotion, termination or retirement' (Chiquita, 2000, p9) – and includes caste, age, sexual orientation, union membership and veteran status, together with the more typical criteria of gender, race, religion or national origin. This coverage is very comprehensive in comparison to other codes (Smith and Feldman, 2003, p21). Chiquita also abides by the equal opportunity provision contained in SA 8000, although the Chiquita code does not contain a specific reference to equal remuneration and promotion of women. Although Chiquita does 'recognize the importance of family in the lives' of employees in their core value of 'respect', it defers to national law rather than specifically mentioning maternity rights and childcare in the code itself (it is therefore part of the code of conduct's compliance with national laws). For example, the Labour Code of Nicaragua states that 'pregnant workers will be entitled to four weeks' leave before childbirth and eight weeks after, with full entitlement to their wage'. In Costa Rica, pregnant women are assigned appropriate tasks, and provided with monthly medical visits and one-hour breastfeeding breaks. Similarly, while the Chiquita code does not make specific commitments to support female employees with policies surrounding reproductive responsibilities, Chapter 13 of the Labour Code obliges employers to provide field workers with a range of services that facilitate combining domestic responsibilities with paid work, such as educational facilities for children and transport to and from work. Finally, sexual harassment is also included in the code provision that covers various harassment situations; Chiquita has one of the very few codes that include this (Smith and Feldman, 2003).

In practice, Chiquita is aware that women face a number of gender-specific concerns and is attempting to tackle some of them. The company has a policy of non-exclusion, equal rights and equal opportunities, and a wide training programme in human rights, including women's rights. In 2004, 13,000 staff members were trained on human rights (including workers and managers). Yet, Jaksch, Chiquita's senior director of corporate responsibility and public affairs, recognizes that there is a gap between company policy and the reality of women's lives, in part stemming from women's subordinate position in several Latin American countries. He explains that 'Our policy is a simple one: to give equal access because what is important is for women to have the training/education needed in order to compete in equal conditions.' However, women still face barriers to promotion and the acquisition of supervisory positions. While Jaksch described a number of positive changes the company has made in the position of women, he also acknowledges that few women occupy supervisory positions, and more remains to be done in this area. This was echoed by a Chiquita manager in Costa Rica who said that, 'Even though Chiquita has been doing an extensive equal opportunity hiring campaign, most of the females apply only for some positions in the packing houses, reducing the opportunities spectrum that can be offered. This behaviour is embedded into local ways of thinking.'

Chiquita also recognizes the problem of sexual harassment and the link between sexual harassment and temporary work. Jaksch described how sexual harassment surfaced as a problem when several cases were identified in the first audits conducted in Guatemala, with the result that two male workers were dismissed. Jaksch emphasizes that sexual harassment is 'not only inadmissible but also illegal', with the consequence that the perpetrator will not only face 'losing [his] job but also a criminal process under Guatemalan legislation. No cases were found according to the 2002 report and all divisions were compliant with SA 8000's provision against sexually abusive behaviour.'

According to Jaksch, a number of other steps have been taken to benefit women, including the provision of transport or help to buy bicycles; the promotion of women with professional qualifications to supervisory positions; and the provision of gender equality in access to housing (with 25 per cent of houses owned by women). Thus, the company has managed to address some of the issues that women consider important in a proactive manner, certainly to a greater extent than those found among independent producers. However, there is still scope for improvement. Chiquita needs to examine whether equal opportunities and freedom from discrimination are being implemented and whether these adequately address the needs of women. It can also be questioned whether women workers wish to work fewer hours to look after children when wages are so low (especially in countries such as Nicaragua); whether local 'culture' is used as an excuse for not promoting women; and whether auditing methodologies fail to identify sexual harassment, as women themselves suggested.

Chiquita speaks about its challenges

A large banana multinational such as Chiquita faces a number of challenges in improving labour conditions, which differ from those in other sectors and also depend upon whether the company produces on its own plantations or purchases bananas from subcontractors/independent producers. According to George Jaksch, the main challenges that the company faces in accomplishing its CSR objectives relate to three areas: internal organizational issues, 'independent' producers and the international banana trade regime

Organizational challenge

As a large multinational, Chiquita must institute a number of organizational changes to effectively implement a CSR strategy. Chiquita's new chief executive officer (CEO), Mexican-born Fernando Aguirre, together with the top management, have promoted the institutionalization of CSR at all levels of the company. According to Jaksch, this shift in organizational philosophy has increased internal motivation and imparted a 'conscience' to the company, thereby fostering improved communication, participation and greater respect for people. This is evident in the number of organizational changes the company has made, including the creation of staff posts for CSR activities; staff training at all levels; the integration of CSR within core business strategy; internal audits; reports highlighting areas of non-compliance; corrective actions; the IUF agreement; and selecting the SA 8000 standard (Werre, 2003).

Jaksch recognizes that financial resources are required to support these internal efforts. However, Chiquita operates in more than 40 countries with 20,500 banana division employees and many more in the subcontracted divisions, and resources are limited. The breadth of Chiquita's operations explains, in part, why some policies that Chiquita has put in place in Costa Rica on their owned plantations are not implemented across second-tier suppliers. The company has yet to extend its policies to independent producers in Nicaragua.

Independent producers

Chiquita insists that CSR objectives should be applied equally in every 'corner' of the company, including among independent producers. In an ILO case study that examines the implementation of the IUF agreement, Riisgaard (2004, p14) argues that 'unsurprisingly, the fieldwork indicated worse conditions on supplier plantations than on plantations owned directly by Chiquita' and those without trade union presence (COLSIBA, 2001, p 43). Yet, while the IUF agreement affords Chiquita the right to withdraw contracts in the event of poor labour conditions and continued supplier non-compliance,[14] such decisions can lead to widespread job losses, particularly in Nicaragua, where Chiquita has subcontracted its entire production. However, workers should not have to choose between poor labour conditions and unemployment. The IUF agreement is particularly important in this regard since 'by covering all Chiquita and supplier plantations in Latin America, [it] can contribute to answering the often real threat of moving production to further exploit differences in labour costs' (Riisgaard, 2004, p17).

However, Chiquita is committed to raising labour standards among its independent producers and intends 'to continue to influence our independent banana growers in Latin America and our joint venture partners in the Philippines to establish target dates for third-party certification to the Better Banana Project and SA 8000 standards' (Chiquita, 2003). Chiquita's 2002 report to the ETI, for example, acknowledged that its CSR efforts were expanding to independent producers and to the development of effective strategies to monitor them (Chiquita, 2002). Jaksch emphasizes that today all growers signing new contracts with Chiquita or renewing contracts are required to obtain Rainforest Alliance certification within a certain time frame, which involves annual independent social and environmental audits of each farm by the non-governmental organizations (NGOs) of the Sustainable Agriculture Network.

Yet, according to Jaksch, the company still faces difficulties in ensuring compliance among independent producers. These difficulties often extend beyond corporate policies to the cultural context of the countries in which Chiquita operates. For example, in many Latin American countries freedom of association is considered a revolutionary act and such sentiments can be hard to overcome. However, Chiquita intends to combine supervision (obligation) with information (communication) in a cooperative way with independent producers, making it known that social and environmental standards are very important for European customers and, hence, for continued business.

The industry

Chiquita also faces specific challenges related to the commercial pressures in the industry, specifically the long-term deterioration in prices and the increasing power of Northern supermarkets, both of which are affecting workers and small banana producers. Indeed, the price wars and buying practices of Northern supermarkets often contribute to poor working conditions as suppliers attempt to deflect market pressures onto workers. As several studies have shown, the buying policies of global retailers often contradict their public commitment to ethical sourcing (Raworth, 2004; Smith et al, 2004). Chiquita contends that supermarkets also have a responsibility to foster decent working conditions on the plantations and packhouses from which they source. Jaksch, for example, calls for the large supermarket chains 'to exercise their power much more, insisting on high standards of corporate responsibility [to all banana companies] because they can drive change very effectively, and they often haven't really begun to awaken to the fact that the power for good that they can exercise is huge'.

Jaksch maintains that a single company like Chiquita cannot institute major changes on its own, but needs to work together with a range of stakeholders. This includes retailers, who are proactive in their CSR strategies – for example, Migros in Switzerland; the Co-operative Group in the UK; the Coop in Italy; and Kesko in Finland – and greatly assist suppliers such as Chiquita. For Chiquita, it also includes the development of alliances with other companies, both within and outside the sector, local authorities, national governments, trade unions, NGOs and international organizations. The company is clear that cooperation is not only more effective at achieving ethical production, but can also encourage the sharing of resources, which is significant in a competitive business climate. While cooperation is not always easy, due to the often conflicting interests of stakeholders, a productive CSR approach depends upon it.

Concluding remarks: Assessing Chiquita's corporate social responsibility policies

There is currently insufficient evidence to draw firm conclusions on Chiquita's overall ethical performance. To date, there is a dearth of independent critical research on the impact of Chiquita's (or other banana multinationals') CSR policies on workers, and even less on the 'improvements' that have been achieved by CSR policies. Nevertheless, this chapter, which compares Chiquita's written commitments and practices with the issues raised by women workers, shows that Chiquita has taken important steps towards conducting business in a socially responsible manner.

While the goodwill of the company is evident, this chapter demonstrates the challenges that are involved in implementing a code of conduct in corporate supply chains. Some of the challenges that Chiquita faces are internal to the organization; others arise from the wider environment. Some develop as a result of problems facing all of society, such as a patriarchal culture, poverty and dependence upon foreign investment. Others arise from a history of labour conflict.

The issues are complex. However, this chapter also points to a number of concrete strategies that any company committed to CSR can put in place. The focus group discussions with women workers, for example, showed that codes could be more effective if there were training for raising awareness amongst workers and managers, better monitoring practices, and an increased commitment from Chiquita and local managers. Similarly, the interviews with Chiquita demonstrated that CSR practices would work better if the company backed its commitment with internal resources and if there were ongoing collaboration with different stakeholders, including local actors such as the monitoring group. Some good practices are the training of workers, working with suppliers and making provisions for women workers.

Bringing different perspectives to this analysis is important. Whereas women workers' views on their labour conditions, and their perspectives and recommendations on codes of conduct, are crucial to any analysis of this kind, the perspective of the companies and the challenges they face are also critical. Given the complexity of the issues involved, the responsibility for achieving a long-term and sustainable change in labour conditions is not Chiquita's alone, but requires a broad stakeholder commitment to socially responsible production. The company and the other actors (including supermarkets, other banana companies, governments, international organizations, trade unions and NGOs) need to work together and engage fully in tackling the wider structural issues that directly contribute to the problems that workers face. Without such broad-based stakeholder engagement, CSR initiatives adopted by multinationals such as Chiquita are unlikely to deliver benefits to workers in the current international political economy.

Notes

1   The views expressed in this chapter are solely the author's and do not reflect the author's involvement with specific organizations. Some of the material comes from fieldwork I conducted during the summer of 2001 as part of an investigation for the New Academy of Business, funded by the UK Department for International Development (DFID), and my doctoral research work at the University of Bristol (Prieto and Bendell, 2002; Prieto, 2004). A shorter related article called 'Corporate social responsibility in Latin America: Chiquita, women banana workers and structural inequalities' was published by the *Journal of Corporate Citizenship* in March 2006. Thanks to the Economic and Social Research Council (ESRC) for funding my post-doctoral research work.

2   The in-depth interview with George Jaksch was conducted by telephone on 24 May 2004. The questionnaire from Costa Rica was received 14 May 2004. At the time, it was not possible to access data directly from Chiquita managers in Nicaragua. I am grateful to George Jaksch for his help in providing information and for his time. Many more thanks go to the women workers who participated in the focus group discussions.

3   The SA 8000 was chosen by Chiquita because it is credible, verifiable, based on international standards and a product of extensive dialogue among stakeholders; it also includes a compliance management system (Zaffa, 2001). For information on SA 8000 see www.sa-intl.org/.

4   It could be argued that these producers are not independent from the transnationals, but rather suppliers in permanent reserve depending upon market demand (Gavilan, undated).

5   Poor working conditions persist despite the existence of legal protections. In 1967, Nicaragua ratified the 1948 Convention No 87 on Freedom of Association; the 1949 Convention No 98 on Collective Bargaining; the 1951 Convention No 100 on Equality of Remuneration; and the 1958 Convention No 111 on the Freedom from Discrimination (Employment and Occupation). In 1981, Nicaragua ratified the 1973 Convention No 138 on Minimum Working Age. These conventions shape relevant aspects of the Nicaraguan Labour Code, updated in 1996.

6   Comprehensive information on the banana trade is available on the United Nations Conference on Trade and Development (UNCTAD) website.

7   These figures are derived from data obtained by UNCTAD and Banana Link, and include other sectors.

8   Bananas are worth UK£750 million each year to the biggest UK supermarkets, second only to petrol and lottery tickets (Ryle, 2003).

9   For example, Chiquita claimed that international market conditions (over-production and low prices) forced it to cut back its order by 180,000 weekly boxes of bananas during ten weeks in 2003, affecting 4000 workers and 15 farms. However, the local association of producers (ANAPROBAN) contends that Chiquita had diverted its purchases to Guatemala, where prices were lower. A few weeks later Chiquita agreed to continue the purchase, but at a price of US$3 a box when the official price set by the government is US$5 (Alvarado, 2003).

10   Jem Bendell's (2001) study of women workers in a Chiquita banana plantation in Costa Rica reports similar labour conditions.

11   Several women in the focus group discussions agreed that unions had not been very attentive to women's concerns.

12   The next Chiquita corporate responsibility report will be published in mid-2006 based on the company's performance in 2003–2005.

13   See Coleman (2002) for an interesting analysis of feminist theory and corporate citizenship.

14   Chiquita de-listed a small number of independent suppliers for lack of cooperation – for example, in Guatemala for a case of child labour (Chiquita, 2002).

References

Alvarado, E. E. (2003) 'Chiquita No Comprará Banano por 10 semanas' ('Chiquita will not buy bananas for ten weeks), *La Nación*, 3 September 2003

Aseprola (undated) *Relatos e Historias de Vida y de Lucha Sindical de Trabajadoras Bananera (Stories About the Life and the Trade Union Fights of Women Banana Workers)* Chapter 5, www.aseprola.org/documents/muj_bananeras.htm, accessed 5 April 2004

Banana Link (2002) 'Wall-Mart: Threat of war in the UK', *Banana Trade News Bulletin*, no 25, March 2002, Norwich, p11

Banana Link (2003a) 'UK supermarket: Leaders in race to the bottom', *Banana Trade News Bulletin*, no 28, July 2003, Norwich, pp1–2

Banana Link (2003b) 'British supermarkets driving the "race to the bottom"', *Banana Trade News Bulletin*, no 29, November 2003, Norwich, pp3

Banana Link (2004) 'An end to discrimination and exploitation? Women fight for a voice at work and in their union', *Union 2 Union: News from the Banana Front*, spring 2004, no 5, pp1, 3–5

Barrientos, S., Bee, A., Matear, A. and Vogel, I. (1999) *Women and Agribusiness – Working Miracles in the Chilean Fruit Export Sector*, Macmillan and St Martin's Press, London

Bendell, J. (2001) 'Towards participatory workplace appraisal: Report from a focus group of women banana workers', *Research Paper*, New Academy of Business

Chiquita (2000) *Code of Conduct*, www.chiquita.com

Chiquita (2002) *Ethical Trading Initiative Report*, www.chiquita.com/corpres/ETI2002.pdf, accessed April 2004

Chiquita (2003) *Sustaining Progress: 2002 Corporate Responsibility Report*, www.chiquita.com

Chiquita, COLSIBA (Latin American Banana Workers' Union Coordination) and IUF (International Union of Food Workers) (2001) *Acuerdo entre la UITA/COLSIBA y Chiquita sobre libertad sindical, las normas laborales mínimas y el empleo en las operaciones bananeras en América Latina* (IUF/COLSIBA and Chiquita agreement on freedom of association, minimum labour standards and employment in Latin American banana operations), www.iufdocuments.org/www/documents/agreements/Chiquita-e.pdf

Coleman, G. (2002) 'Gender, power and post-structuralism in corporate citizenship: A personal perspective on theory and change', *Journal of Corporate Citizenship*, spring, pp17–25

COLSIBA (Latin American Banana Workers' Union Coordination) (2001) *Diagnostico Participativo con Enfoque de Género sobre Condiciones Sociales, Económicas, Laborales y Organizativas de las Mujeres Trabajadoras Bananeras*, Mimeo, Honduras, www.colsiba.org/mujertrabajadora.htm

Dolan, C., Opondo, M. and Smith, S. (2003) 'Gender, rights and participation in the Kenya cut flower industry', *Natural Resources Institute Report*, no 2768, SSR Project no R8077 2002-4, University of Greenwich, UK

Enloe, C. (2000) *Banana, Beaches and Bases: Making Feminist Sense of International Politics*, University of California Press, Berkeley

Ethical Performance (2001) *Chiquita: Sustainable Development, Best Practice Case Study*, autumn 2001, www.ethicalperformance.com, accessed on 5 April 2004

Gavilan, J. (undated) 'El banano de exportación en Nicaragua', http://bananasite.galeon.com/banano.html, accessed April 2004

Harris, C. (2003) 'Profile: Iris Munguia, union coordinator', *Big Issue in Scotland*, 4–10 December, pp32–33

ILO (International Labour Organization) (2004) *Global Employment Trends for Women 2004: Key Indicators of the Labour Market*, ILO, Geneva

Mokhiber, R. and Wheat, A. (1995) 'Shameless: 1995s ten worst corporations', *Multinational Monitor*, vol 16, no 12, http://multinationalmonitor.org/hyper/mm1295.04. html, accessed April 2004

Naranjo, A. V. (1999) *Health Risks for Women Banana Workers*, Mimeo, Aseprola, San José

Pearson, R. and Seyfang, G. (2002) 'I'll tell you what I want: Women workers and codes of conduct', in Jenkins, R., Pearson, R. and Seyfang, G. (eds) *Corporate Responsibility and Labour Rights: Codes of Conduct in the Global Economy*, Earthscan, London

Prieto, M. (2004) 'Is there anyone listening? Women workers in factories in Central America, and corporate codes of conduct', *Development*, vol 47, no 3, pp101–105

Prieto, M. (2006) 'Corporate social responsibility in Latin America: Chiquita, women banana workers and structural inequalities', *Journal of Corporate Citizenship*, vol 21, Spring

Prieto, M. and Bendell, J. (2002) 'If you want to help us then start listening to us! From factories and plantations women speak out about corporate responsibility', *Occasional Paper*, www.new-academy.ac.uk/research/gendercodesauditing/index. htm

Raworth, K. (2004) *Trading Away Our Rights: Women Working in Global Supply Chains*, Oxfam, Oxford

Riisgaard, L (2004) 'The IUF/COSILBA – Chiquita framework agreement: A case study', *Working Paper No* 94, ILO, www.ilo.org/public/english/employment/multi/ download/wp94.pdf, accessed December 2004

Ryle, S. (2003) 'Banana war leaves the Caribbean a casualty', *Corporate Accountability, Observer Newspaper Special*, 24 November 2003

Smith, S., Auret, D., Barrientos, S., Dolan, C., Kleinbooi, K., Njobvu, C., Opondo, M. and Tallontire, A. (2004) 'Ethical trade in African horticulture: Gender, rights and participation', *IDS Working Paper*, no 223, IDS, Brighton

Smith, G. and Feldman, D. (2003) *Company Codes of Conduct and International Standards: An Analytical Comparison*, World Bank and International Finance Corporation, Washington, DC

STITCH (Support Team International of Textileras, Organizers for Labour Justice) (undated) *The Truth on Your Table: Facts about Women Workers in the Banana Industry*, STITCH, www.stitchonline.org/archives/organizer.asp

UNCTAD (United Nations Conference on Trade and Development) (undated) *Info Comm: Market Information in the Commodities Areas*, http://r0.unctad.org/infocomm/ anglais/banana/sitemap.htm, accessed 16 December 2005

Vorley, B. (2003) 'Food, Inc.: Corporate concentration from farm to consumer', UK Food Group', www.ukfg.org.uk, accessed 5 April 2004

Werre, M. (2003) 'Implementing corporate responsibility – the Chiquita case', *Journal of Business Ethics*, vol 44, May, pp247–260

Zaffa, J. (2001) 'Corporate responsibility at Chiquita', Presentation to Banana Link Conference on Voluntary Social Standards in the Banana Industry, 27 September 2001, London

7

The Gangmaster System in the UK: Perspective of a Trade Unionist

Don Pollard

Introduction

For over a century, the gangmaster system has been an important part of UK agricultural and food production, and now provides over half of the labour needed for planting, cultivating, harvesting and packaging of UK fruit and vegetables. Gangmasters – labour providers who recruit and organize groups of people to work on farms and in packhouses – are involved in a legal activity that provides a flexible labour force needed to meet the seasonal production and market demands of the food industry. However, within this framework of legal and essential labour provision there are, and always have been, serious abuses of workers' rights.

In the past, the gangmaster system was localized and workers were recruited from the rural areas surrounding farms and packhouses. Worker abuses, such as low pay, long hours and poor conditions, were generally tolerated as rural people, especially women and young people, faced few alternative prospects for work and income. However, over the last few decades the agricultural industry has moved away from this model of localized labour towards an immigrant-based gangmaster system. Many of these immigrants are working illegally in the UK and their illegal status places them in a particularly precarious position. In an industry where union organization is rare, gangworkers have virtually no trade union representation to protect their interests and improve their conditions. Furthermore, few are afforded the protection of existing agricultural and employment law, either due to a lack of awareness and/or intimidation by unscrupulous gangmasters. People trafficking, financial exploitation, health and safety violations, housing abuses, tax evasion, physical intimidation and other abuses of workers' rights are endemic in the gangmaster system, and are faced daily by UK and migrant gangworkers – legal and illegal.

This chapter traces the evolution of the UK gangmaster system and outlines the changes in agricultural production, organization and food marketing that have precipitated a shift away from localized seasonal labour to a reliance on year-round immigrant workers on large-scale farms, greenhouses and packing stations. It describes the abuses associated with the current system and identifies the ways in which unscrupulous gangmasters have been able to evade the law. The chapter focuses specifically on measures that have recently been taken to eradicate abuses, including trade union and parliamentary campaigning, which have led to the establishment of Operation Gangmaster, the Ethical Trading Initiative (ETI) Temporary Labour (Gangmaster Working Group CTLWG), and the introduction of the Gangmaster Act in 2004. The chapter concludes with a discussion of the long-term potential for reducing worker and tax abuses, and increasing ethical production and purchasing in UK agriculture.

Supermarkets and changes in the food system

Over the past 20 years, there have been major changes in the organization of agricultural production and food marketing in the UK, which underlie the current transformations occurring in the gangmaster system. Most notably, among these has been the growth in the dominance of UK supermarkets and their increasing control over the supply chains of food products, particularly fresh fruits and vegetables. With retail concentration at unprecedented levels, supermarkets now exercise considerable control over which products are produced, and how they are grown, packed and delivered to stores throughout the UK. Their requirements have had a dramatic impact on growers and packers, whose business is increasingly governed by the demands of their super-market customers, demands that have accelerated the trend towards large-scale growers and packhouses, and centralized depots for food distribution. Supermarkets have responded to public criticisms surrounding their market power by claiming that they are simply responding to customer preferences for better food at cheaper prices. But there is a human cost to cheap prices; while supermarkets may be providing UK consumers with quality and variety at reasonable prices, their suppliers are forced to cut their production costs or go out of business altogether. When suppliers are forced to trim production costs, however, they look to labour to provide their cost savings. This has meant a decline in direct employment and a concomitant shift to the use of gangmasters.

A second factor contributing to an increased reliance on the gangmaster system is the ordering practices of supermarkets. Computerized ordering, based on hour-by-hour sales returns from checkout counters, means that supermarkets can, and do, change orders at very short notice. Their suppliers of fresh produce must then alter their production to meet these changing orders. This, in turn, means that workers must be brought in when orders are increased or sent home when the orders are cut. The gangmaster system is best placed to meet these flexible demands; but gangworkers themselves are the victims of this employment instability.

The supermarkets often proclaim that they have a partnership with their suppliers. However, it is an unbalanced one, even for large producers who have expanded and profited from their relationships with the supermarkets. This asymmetry is evident in a number of practices. During annual order discussions, for example, supermarkets, not their suppliers, determine the type and quantity of crops to be produced. Supermarkets often impose a levy on growers and packhouses for sale promotions and insist on seven-day deliveries, while growers and packhouses are paid on the basis of market prices rather than production costs. Supermarkets carry out quality control of production and packing, but insist that growers bear the cost of social audits. These practices occur in a context in which the profit margins realized from supermarkets' sales of fresh fruit and vegetables are the highest in Europe.

However, accompanying the commercial changes driven by supermarkets is increasing public attention on the ethical behaviour of corporations. This growing consumer concern has prompted UK supermarkets to adopt a range of ethical activities and programmes. Codes of labour practice governing the operations of supermarkets and their suppliers are now common. Most of the major supermarkets are also members of the Ethical Trading Initiative (ETI) whose Base Code sets out basic workers rights (see Chapter 1). In addition, fair trade-labelled products are increasingly found on supermarket shelves. Yet, despite these trends, supermarkets remain locked in a competitive battle with their rivals and continue to place intense pressures on suppliers to reduce costs. The outcome is a gangmaster system characterized by worker exploitation and tax evasion.[1]

Changes in the system: The gangmasters

Gangmasters have developed from small-scale, locally based businesses employing a minibus of workers (about 15 to 20) to a system where 100-plus workers are on their books (and even more operating 'off' their books). Some have over 1000 gangworkers and financial turnovers of several million UK pounds. Many now call themselves employment agencies or labour providers and have expanded their operations into non-agricultural labour.

The importance of gangmasters has increased during recent years as unemployment has declined in rural areas. Many UK residents are no longer willing to work for low pay and long hours, and will not tolerate the dirty conditions found on many farms and in packhouses. To meet this labour gap, gangmasters have expanded their use of immigrant labour. Some of these immigrants are working legally through the Seasonal Agricultural Workers Scheme (SAWS), the Commonwealth Working Holidays Visas and a small handful of official government-approved programmes. However, much migrant labour is illegal. This has provided an opportunity for less scrupulous and even criminally connected gangmasters to take advantage of precarious workers whose legal status forces them to accept whatever conditions are imposed on them. Having exploited UK workers for years, unscrupulous gangmasters have found a new group to profit from through abusive labour practices.

Yet, many of the exploitative practices associated with gangmasters relate to the commercial environment described above. Supermarkets, for example, exert intense pressure on their suppliers to reduce prices, and they in turn deflect this pressure onto their labour providers, the gangmasters, on a take-it-or-leave-it basis. The profit margins demanded by farmers/packhouses can make it difficult for gangmasters to meet their legitimate employment costs. As a consequence, many unscrupulous gangmasters flout laws on pay, conditions of work and health and safety. For instance, gangmasters often recover money from workers illegally or engage in unethical practices, such as taking excessive deductions from workers' pay for accommodation, transport, interest on loans, clothing and work-related equipment, and may levy a vague charge for administration. Those gangmasters who wish to comply with employment laws such as some of the older, smaller gangmasters, often find it difficult to compete. Many cut corners to survive or become subcontractors for larger gangmasters, while others have gone out of business altogether. In fact, subcontracting (where one gangmaster has an agreement with a farmer/packhouse to provide a certain number of workers but uses another gangmaster to supply those workers) has become a major feature of the current gangmaster system. It is not unusual to find a main gangmaster subcontracting with two or more sub-gangmasters. In some cases, large gangmasters use subcontractors to specifically deal with the illegal aspects of the business, such as the use of migrant workers. Such double or off-the-book recording obviates a gangmaster's direct responsibility for any violations of the law, allowing a gangmaster to present a legitimate face to authorities while skirting the law through clandestine means.[2] Frequent changes in the names of gangmaster companies, addresses and directors further obscures the situation for enforcement agencies.

Abuses in the gangmaster system

There are many existing UK laws relating to employment rights, health and safety, housing and other matters concerning worker welfare. There are also immigration laws that stipulate the right to live and work in the UK. While it is individuals who break these laws, it is those who arrange their travel, work and accommodation who most profit. Individuals are drawn into a system of exploitation developed by gangmasters to further their own interests while protecting them from the law. The following describes the types of legal violations perpetuated by unscrupulous gangmasters:

- *Non-compliance with statutory wage rates.* In the UK there are currently two wage rates that have the force of law: the National Minimum Wage (NMW) and the Agricultural Wages Board (AWB) Order. The NMW applies to all workers and is set at an hourly rate on an annual basis. The

AWB Order applies only to those working in defined agricultural work and establishes both hourly and overtime rates. The AWB contains different rates, ranging from harvest workers paid at the NMW rate to regular workers paid at a higher standard worker rate, to even higher rates for skilled and supervisory workers. The AWB Order also establishes the daily and weekly working limits after which overtime, holiday entitlement, sick pay and accommodation must be paid. These rates and conditions are the result of annual negotiations between the Transport and General Workers' Union, the National Farmers' Union and independent members of the AWB. Many gangmasters fail to consider the AWB pay and conditions even where gangworkers are covered by it and, at best, pay the NMW rate. Gangmasters often pay on a piece rate basis, which can result in workers being paid below the minimum rates. Many regular gangworkers, mostly women, do not receive their full entitlement to holiday pay and time, the special Agricultural Sick Pay Scheme and many other provisions of the AWB Order. In essence, the laws, especially the AWB Order, are often not applied in the case of gangworkers, and many workers are not aware of these rates or of their rights to receive them.

- *Unlawful deductions from wages.* The only legal deductions from wages without the agreement of the worker are for income and national insurance taxes. However, it is common practice for gangmasters to deduct money for accommodation, transport, work clothing and equipment, interest on loans, administrative charges and other unexplained reasons without a worker's permission. Many gangworkers do not receive payslips outlining the details of these deductions. The result is that workers are often remunerated for a small fraction of what they should earn for the hours they work. It is not uncommon for a gangworker to actually owe money to the gangmaster at the end of the week if there has been little work, but the deductions for accommodation, etc., continue.

- *Undeclared labour.* Many gangmasters offer workers the choice to be on or off their books. If workers choose the latter, they are usually paid less than registered workers, but no deductions are taken for taxes and they receive 'cash in hand'. This arrangement suits the gangmaster since it lowers his wage bill. But it also suits those who are working illegally, either because they are in receipt of state benefits through other means or are working illegally in the UK.

- *Health and safety violations.* There have been many examples of accidents, some fatal, involving gangworkers as a result of health and safety violations.[3] Few gangworkers are given safety training even when they are using potentially dangerous machinery and chemicals. This is especially true when the gangworker does not speak or understand English very well. Health and safety regulations are not applied in many situations in which gangworkers have to operate. Drinking water and toilets are often not adequate or are not provided at all, especially in field situations where gangworkers work in hot and dirty conditions. The buses and/or vans used for the transport of gangworkers are often unsafe and are sometimes driven by people without proper experience and qualifications. Long hours and continuous days worked are a serious health risk. Excessive work time is sometimes the result of pressure from the gangmaster. But workers themselves may also decide to work long hours, especially migrant workers who are desperately paying off their debts and saving money to send home to their families.

- *Use of migrant labour.* Although all gangworkers are vulnerable to the abuses described above, migrant workers face a higher likelihood of labour violations, especially if they are working in the UK illegally. For many migrant gangworkers, their exploitation begins before they leave their home countries, continues while they are working in the UK and sometimes persists after they return home. Spurious employment and travel 'agents', who are often linked to criminal organizations and/or have direct connections with UK gangmasters, arrange jobs, visas and accommodation for people desperately seeking work in the European Union (EU). Since most of these people are not able to finance these arrangements themselves, they find themselves in debt to agents at extortionate interest rates. In some cases, agents provide false passports, visas and work permits; however, these visas are often ordinary tourist ones or for students enrolled in bogus English language schools. While tourists are not allowed to work, students are permitted to work up to 20 hours a week. All migrants who are then employed full time are working illegally. This, of course, places migrant gangworkers in a highly precarious situation. Yet fear prevents them and those who are aware of such illegal practices from speaking out. Gangworkers know that speaking out can result in an increase in the daily verbal abuses that they face, possible physical intimidation and attack, the loss of their jobs and potential deportation.

However, for the gangmaster, there is little risk entailed in hiring an illegal worker. If a person is found to be working illegally in the UK, for example, they are deported, but the gangmaster who profited from their cheap labour is seldom prosecuted. In the few cases where gangmasters have been convicted, their fines have been less than the income they earned from their gangworkers. There is thus little incentive for a gangmaster to curb illicit practice. Even the recent enlargement of the EU, enabling people from ten Eastern European countries to work legally in the UK, has not eradicated abuses in the system. Despite their legal status, many migrant workers continue to lack information on UK employment law and are easily denied their basic employment and human rights.

The reality of the gangmasters

It should be emphasized that to be a gangmaster is perfectly legal and the service they provide (a cheap and flexible workforce) is essential for the efficient running of the agricultural and food industries. Many gangmasters do operate legitimate businesses and respect the rights and welfare of their workers. But, as noted above, the gangmaster system has become a very competitive business and price pressures from supermarkets are passed from growers and packhouses to labour providers. Gangmasters, like every other business, are out to make a profit. Unfortunately, many turn to illegal means to ensure that profit.

Even where a farmer or packhouse pays the gangmaster the statutory rate for the workers he provides, there is no guarantee that the gangworkers will actually receive that amount. As noted, the absence or inadequacy of employment contracts and payslips is common, preventing workers from having an awareness of their pay and authorities from knowing whether appropriate regulations are met. Furthermore, gangmasters use the money they earn through the provision of accommodation, transport and loans as an additional income source, and often charge prices that far exceed their costs. In the worst cases, people trafficking, drugs and money laundering have been connected with gangmasters.

Even where the law is meant to apply to gangworkers, it does not necessarily extend protections to illegal migrant workers. In July 2004, a request was made to the UK Department for Environment, Food and Rural Affairs (DEFRA) for an investigation into the enforcement of the AWB Order concerning gangworkers who were not paid the proper AWB rates. In response, a DEFRA official wrote a letter stating:

> *Advice received from the Department for Trade and Industry holds that workers who enter the country illegally do not benefit from many protections found in UK employment law. As a consequence, the Agricultural Wages Team (AWT) has no powers to enforce the provisions of the Agricultural Wages Order (AWO) in respect of illegal workers.*

This effectively meant that gangmasters (or anyone else) who used illegal migrant workers were free to pay them what they wished since these workers were not protected by law. Yet, by placing illegal migrant workers outside the law, gangmasters were given a legal green light to maintain abusive practices.

'Something must be done' and has

One of the strange things about gangmaster abuses has been that everybody involved in the food industry, from workers to trade unions, to farmers to packhouses, to supermarkets, as well as enforcement agencies, were aware that the situation was not only bad, but was getting steadily worse, and that something needed to be done. Yet, given the extent of stakeholder concern, why did action take so long to materialize? The reasons are multifaceted. On the one hand, competing legislative priorities diminished the importance of a legislative response to the gangmaster system. Where initial action was taken by enforcement agencies acting on behalf of government ministries, both Tory and Labour, they sought to institute change through voluntary best practice rather than through legislative means. Furthermore, agricultural workers, and especially gangworkers, were typically reluctant to join trade unions or act collectively due to the prospect of employer intimidation. Trade unions themselves found it difficult to organize gangworkers who were temporary, widely dispersed and often faced language problems. Farmers and packhouses were under financial pressure from the supermarkets and generally did not consider gangworkers as part of their own workforce. Supermarkets were doubly removed from the problems as they did not employ the gangworkers themselves, nor, in fact, did their suppliers. In this context, where no one party assumed responsibility for worker exploitation, gangmasters were free to operate without restraint.

This is not to imply, however, that the issue of gangmasers never attracted government attention. Prior to World War II, the National Union of Agricultural and Allied Workers and a handful of MPs debated gangmaster abuses and advocated concerted actions at a number of annual conferences. In September 1995, the Agricultural Workers' Section of the Transport and General Workers' Union (T&GWU), which had merged with the old National Union of Allied and Agricultural Workers (NUAAW), raised the issue of gangmaster abuse at an international conference of the European Federation of Agricultural Workers (EFA), and the EFA subsequently approached the European Commission to fund a research project on the topic. Following the project's approval, the Office for European Social Research (ORSEU) was commissioned to organize and analyse a survey to be carried out in the UK, France, The Netherlands, Germany, Italy and Spain. The T&GWU undertook to do the research in the UK, mainly in East Anglia and Lincolnshire, launching one of the first attempts to identify the issues facing gangworkers. The report, which was completed by the autumn of 1997,[4] highlighted the state of the gangmaster system in the UK, documenting the nature of abuses confronting workers and the difficulties entailed in enforcing laws to eradicate them. In particular, the ORSEU report emphasized the need for a statutory registration scheme for gangmasters, a long-time demand of the union. At the same time, meetings were held between representatives of the National Farmers' Union (NFU), the T&GWU and officials from the Ministry of Agriculture (MAFF) to examine gangmaster issues, leading to the publication of the Working Party on Agricultural Gangmasters' report in December 1997.[5]

Operation Gangmaster

The culmination of these activities was the launching of Operation Gangmaster in June 1998. The operation was a welcome example of interdepartmental cooperation that brought together nine ministries and agencies involved in enforcing laws relating to gangmaster abuses. It also developed a number of new strategies to combat gangmaster abuse, including highly publicized blitzes of farms and packhouses by enforcement agencies, and the publication and distribution of leaflets by MAFF describing the rights of workers[6] and the responsibilities of employers. The Lincolnshire area was selected for the first trials of the operation as it is a region where many gangmasters work, and Operation Gangmaster teams were subsequently established in five other regions of the UK.

Despite its promising start, an early analysis of Operation Gangmaster revealed a number of shortcomings, including a lack of resources allocated to participating enforcement agencies, and a lack of cooperation among farmers and packhouses who were lax in checking the credentials of their gangmasters or the workforce supplied.[7] Furthermore, the report noted that an 'awareness of Operation Gangmaster did not seem to have filtered through to the individual multiple retailers [supermarkets] ... though they are well aware of the problems'.[8] Some of the growers were even more critical of supermarkets. One said: 'The supermarkets are very commercial... They want us to pay low wages, then they can push the price down further.' Another said: 'Supermarkets are two-faced ... they are only interested in price.'[9] More bluntly, a grower charged: 'The supermarkets are completely ruthless.' A strong recommendation in both the Executive summary and Conclusion of the 1999 Operation Gangmaster report was that: 'the only effective sanction will be compulsory registration of gangmasters'.[10]

Voluntary initiatives

In an era of corporate social responsibility and ethical trading, codes of practice governing worker welfare have proliferated. The gangmaster system, too, has been affected by this trend and a number of organizations have produced codes of practice. These include the National Farmers' Union, the Fresh Produce Consortium (FPC), individual supermarkets and even gangmasters themselves via the Association of Labour Providers (ALP). These codes set out the responsibilities and procedures required to comply with the law on gangmaster matters. However, despite the growth in codes, compliance remains voluntary and is not usually independently monitored or statutorily enforced. As the Executive summary and Conclusion of the 1999 Operation Gangmaster report stated: 'the current NFU code is toothless'.[11] Furthermore, the situation did not improve with the creation of the UK Department for Environment, Food and Rural Affairs (DEFRA), which replaced MAFF. According to a DEFRA report published in September 2003: 'Operation Gangmaster remains a woefully inadequate response to the complex enforcement issues arising from the illegal activities or gangmasters' and 'it is unrealistic to expect the voluntary codes to prevent widespread illegal activities by gangmasters'.[12] The problems with voluntary initiatives were highlighted by Sainsbury's experience, the first supermarket to publicly address concern about gangmaster abuse at their Gangmaster Conference in January 2001. The supermarket brought together senior company officials, their suppliers, trade unionists, non-governmental organizations (NGOs), enforcement agency personnel and gangmasters to increase the understanding of gangmaster issues and to orchestrate a concerted response. Yet, while supermarkets have developed company codes of practice,[13] we know little of how they respond to cases where suppliers are found using gangmasters engaged in illegal practices.

A central difficulty with voluntary initiatives relates to cost: who pays for, monitors and ensures that supermarket guides/codes are carried out. Although the supermarkets are very experienced in conducting audits that evaluate the quality, environmental, and health and safety aspects of their products, they are less experienced in conducting social audits on the labour conditions of their suppliers. Social audits are meant to assess a number of issues, including whether minimum conditions are being met at the place of work, accommodation, wages, deductions and other matters affecting the welfare of gangworkers. These issues can be difficult to audit and many of the audit companies used by supermarkets have not been effective. This has prompted Asda, for example, to employ an auditing company with more experience on social issues.

As noted, many supermarkets are also members of the Ethical Trading Initiative (ETI), and apply its Base Code as a minimum within their own company codes of practice (see Chapter 1). Although the ETI initially focused on working conditions in developing countries through its pilots and working groups, in 2002, the ETI focused attention on the gangmaster issue in the UK, where all the stakeholders in the agricultural and food industry felt that there were growing problems with worker exploitation. This was the first ETI project directly involving a UK issue. It organized regional seminars in Kent in April and in Lincolnshire in May 2002 that were well attended by representatives from the supermarkets, growers, trade unions, enforcement agencies and others engaged in the problems of the gangmaster system. These seminars produced a consensus that little effective action had been taken to address the worsening gangmaster situation. Following the seminars, a Gangmaster Working Group (later called TLWG) was formed consisting of the ETI acting as the secretariat with representation from the supermarkets, the T&GWU, the Trade Union Congress (TUC), a gangmaster, enforcement agencies and civil servants led by DEFRA. The working group held regular meetings and proposed statutory action to establish a registration and licensing system for gangmasters.

Following meetings with ministers from DEFRA and the Home Office, the ETI Gangmaster Working Group agreed that, in addition to pressing for statutory action, it would support voluntary proposals to improve the situation. Amongst the proposals it supported was the creation of a gangmaster organization, the Association of Labour Providers (ALP), which would bring together legitimate and ethical gangmasters to set up a code of practice. Compliance with the code would be voluntary; but the ALP would supervise its members and labour users (i.e. farmers and packhouses), and thereby provide supermarkets with assurances that gangmasters were operating in conformance with the ETI Base Code, the ALP Code of Practice and all UK laws relating to employment and business practices.

In 2004, the ETI Gangmaster Working Group also initiated a pilot auditing project, drawing on different professional audit companies to audit a limited number of gangmasters in the Spalding area of Lincolnshire, a main UK horticultural area with extensive use of gangmasters. The supermarkets were asked to have one of their suppliers identify a gangmaster they used to take part in audits, who were shadowed by members of the ETI working group.

The results were troubling. Even amongst gangmasters who had agreed to be audited, a number of code violations were detected, including widespread abuses of workers' rights and health and safety regulations; inadequate and expensive accommodation provision; and questionable or outright illegal tax and business practices.[14]

The Gangmaster Private Members Bill – Jim Sheridan, MP

In 2003, the Transport and General Workers' Union, frustrated with the lack of progress, decided to support Jim Sheridan, a Labour MP from Scotland, to introduce a private members bill on the registration of gangmasters. While the union had long campaigned for statutory action, the promotion of a private members bill, which lacks government backing, reflected a desperate attempt to push the legislative agenda forward. On 7 January 2004, Sheridan introduced his Gangmaster Bill in Parliament with a second reading scheduled for 27 February. This is the stage where private members bills usually fail due to a lack of support or opposition. Yet, external circumstances intervened to change the course of events. Tragically, on 5 February 21 Chinese cockle pickers were drowned at Morecambe Bay in Lancashire. It was disclosed that they were employed by a gangmaster, and most of them were working in the UK illegally. They had been housed in appallingly crowded conditions and were likely paid below NMW rates. Their deaths, a visible outcome of the gangmaster system, thrust the issue into the media spotlight and drove gangmasters to the top of the political agenda. This tragic turn of events accelerated the process of legislative action, a process that typically requires two years to bear fruit. The Home Secretary stated that the government would back Sheridan's bill and move forward on it quickly. The remarkable conversion of the government, MPs, the media and the general public was made at the highest price for those killed at Morecambe Bay. The Gangmaster Bill received Royal Assent on 8 July 2004. While at the time of writing it awaited important secondary legislation, it nonetheless signifies a landmark in the history of gangmasters.

The future

The accession of the Eastern European states into the European Union means that many immigrant workers no longer experience the fear and exploitation associated with illegal work. However, the abuse of vulnerable workers, such as migrant workers, including women and young people, remains a constant feature of the employment situation. As this chapter has described, considerable efforts have been made to eliminate gangmaster exploitation. Codes of practice, while weak in their capacity to mitigate abusive practices, played an important role in focusing the energies of people and organizations on what was wrong and on devising ways to right those wrongs. However, it is the Gangmaster Act that marks the first real step towards controlling those who benefit from the exploitation of workers. The Gangmaster Act aims to identify and hold gangmasters accountable, including subcontractors, and will also make farmers and packhouses responsible for the labour provided to them by gangmasters. Yet, past experience has shown that unscrupulous gangmasters are adept at avoiding legal obligations by their Hydra-headed changes of names, addresses, directors and the use of subcontractors. The new law will be a help to curb this; but it will not provide a total solution to the problems of worker exploitation.

The struggle to eliminate gangmaster abuses has not ended with the passing of the Gangmaster Act. Two recent proposals have threatened the effectiveness of the act. DEFRA has suggested that the new act should not cover second-stage food processing. This would mean that over 200,000 workers in packing and processing would not be protected. Also recently, the Better Regulations Executive suggested that audits of gangmasters applying for licences under the act should only be made of those who are deemed 'risky'. This suggestion flies in the face of all those involved in the issue – retailers, farmers and trade unions – that *all* gangmasters must be inspected before they are granted a licence.

In addition, there is one important weakness in the act; it does not cover the supermarkets who purchase produce. In fact, the purchasing system that produces the pressure to reduce prices paid for fresh fruit and vegetables remains intact. The supermarkets must make their claim to ethical sourcing a reality. Low prices paid to producers may make food cheaper to the consumer; but the real cost is paid by workers facing employment and human rights abuses. The future for a protected workforce is trade union organization, effective legislation and campaigning for human rights. The disgrace of the current gangmaster system has too long been a part of British history. This stain must now be removed.

Notes

1   Many gangmasters avoid their tax responsibilities. Their failure to pay income and national insurance taxes and VAT is an abuse of all taxpayers as the government loses this income.

2   The Fresh Produce Consortium commissioned a report from the Agricultural Investigation Team, part of the Department of Works and Pensions, in June 2001. It highlighted the problem of subcontracting as a means of main gangmasters avoiding legal responsibility, and of suppliers turning a blind eye to gangmaster abuses in order to fulfil supermarket contracts. The report stated that: 'It is now time for legislation to back up our efforts and help us to control gangmasters in the future' (AIT, 2001, p8).

3   One example was an accident in July 2003, when a minibus carrying immigrant gangworkers in the Vale of Evesham was struck by a train on an isolated level crossing, killing three workers. It is believed that the driver of the minibus could not read the English warning sign and that he had not been given any training about the dangers of the crossing.

4   The ORSEU 2000 report was entitled *Undeclared Work in Agriculture* and was published in November 1997. It was supervised by Jean-Pierre Yonnet with the financial support of the Commission of the European Union Directorate-General V, Employment, Industrial Relations and Social Affairs.

5   The report of the Working Party on Agricultural Gangmasters was entitled *Report of the Interdepartmental Working Party on Agricultural Gangmasters* and was produced by MAFF in December 1997.
6   This was translated into Russian and Polish for the main group of gangmasters involved in the UK at that time.
7   The analysis of Operation Gangmaster was entitled *Economic Evaluation of Operation Gangmaster* and was produced by Produce Studies Limited for MAFF in April 1999, Reference PSL 9898/HWB.
8   MAFF (1999) *Economic Evaluation of Operation Gangmaster*, para 7.4.1, p17.
9   MAFF (1999) *Economic Evaluation of Operation Gangmaster*, para 8.7.3, p42.
10   MAFF (1999) *Economic Evaluation of Operation Gangmaster*, Section (e), pv.
11   MAFF (1999) *Economic Evaluation of Operation Gangmaster*, Section (e), pv.
12   Environmental Food and Rural Affairs Committee (2003, Recommendations 6, 8 and 9, pp5–6).
13   Waitrose also produced its own *Labour Provider/Users Guides*.
14   See Lawrence (2004). Her article in the *Guardian* cited one case, in particular, where the South African workers of a gangmaster were the victims of a whole host of employment and housing abuses. These workers had come to the UK as part of the Commonwealth Working Holidays Scheme and were, therefore, working here legally. Their visas to the UK had been arranged by a travel company in Pretoria with whom the gangmaster had an arrangement. The travel agency provided travel loans, but at 100 per cent interest charges and the workers had to sign an agreement that they would not leave the employment of the gangmaster until the loan was paid off. This agreement also said that legal action would be taken against whoever had given a reference to the worker, usually their parents. Repayment of the loan would be taken from the workers' wages until the full amount was paid. This is a form of bonded labour. Once they had arrived in the UK most of the workers were provided with overcrowded accommodation by the gangmaster. They were paid below the NMW, were discouraged from obtaining national insurance numbers, and had deductions made for transport, rent, loan repayments and sometimes other unspecified charges. They were often required to work long hours and continuous days. When these violations were put to the gangmaster, who was a member of the new ALP, he claimed that he did not employ these workers as he had subcontracted the work to another gangmaster and, therefore, was not liable for any offence.

References

AIT (Agricultural Investigation Team) (2001) *Report for the Fresh Produce Consortium*, AIT, Spalding

Environmental Food and Rural Affairs Committee (2003) *Gangmasters: Government Reply to the Committee's Report: First Special*, 10 December 2003, HC 122, Stationery Office, London

Lawrence, F. (2004) 'Migrants in bonded labour trap', *Guardian*, 29 March 2004

MAFF (UK Ministry of Agriculture) (1999) *Economic Evaluation of Operation Gangmaster*, Produce Studies Limited, Reference PSL 9898/HWB, London

ORSEU (Office for European Social Research) (1997) *Undeclared Work in Agriculture*, University of Lille, Lille

Working Party on Agricultural Gangmasters (1997) *Report of the Interdepartmental Working Party on Agricultural Gangmasters*, MAFF, London

8

Participatory Social Auditing: Developing a Worker-focused Approach

Diana Auret and Stephanie Barrientos

Introduction[1]

Compliance with codes of labour practice has increasingly become a condition of global supply for producers supplying some large retailers and food companies. Codes, which arose in response to civil society campaigns against retailers and brands, have become an important mechanism for staving off 'risk' to the brand image as a result of adverse publicity for poor supply chain labour conditions. They set employment standards that suppliers must meet if they wish to supply certain companies. Verifying compliance with company codes has spawned a growth industry of organizations that carry out social audits of suppliers against different buying company codes of labour practice. These normally involve a 'snapshot' visit by a compliance officer employed by the retailer or buyer, or an external professional auditor, to a farm or producer to monitor compliance against the buyer's code. However, increasingly, criticism is being raised of snapshot audits. They tend to focus on formal management compliance rather than helping to support genuine improvement in workers' rights. They tend to pick up 'visible' issues, such as health and safety, but often fail to pick up issues that are not easily verified by company records or physical inspection, such as freedom of association or discrimination. They are often insensitive to issues of concern to women workers, or insecure workers such as temporary, migrant and contract labour. This is important given the large number of such workers increasingly employed in the global food system (see Chapter 5).

A participatory approach to auditing codes of labour practice adopts a different perspective than more compliance-focused 'snapshot' social auditing. It puts greater emphasis on the involvement of workers and workers' organizations in the process of code implementation and assessment. It is based on developing partnerships between different actors (companies, trade unions, non-governmental organizations – NGOs – and government) in developing a locally sustainable approach to improving working conditions. Because of this, a participatory approach is better equipped to uncover and thus address more complex issues, such as gender discrimination and sexual harassment. These are less 'visible' issues that are unlikely to be resolved through a simple compliance approach and are more likely to be experienced by insecure non-permanent workers, who are often women. The goal of a participatory approach is a process of awareness creation and improvement that is more sensitive to gender issues, and the conditions of insecure and vulnerable workers.

A participatory approach can be developed at different levels. At a minimum, it involves the use of participatory tools in the process of social auditing to ensure that the views and voices of workers, especially women workers, are captured in an audit. More genuine participation by workers also requires the involvement of workers' representatives or shop stewards at site level, and sector trade unions and NGOs, both in awareness creation and in the auditing process. At its broadest level, it involves the development of local 'multi-stakeholder' initiatives. These bring companies, trade unions and NGOs together with government to form an independent body able to oversee the implementation and monitoring of a locally relevant code of labour practice. Although a participatory approach faces many challenges, it represents a shift away from a top-down management compliance orientation to the greater empowerment of workers and their representative organizations as an essential part of the process of improving working conditions.

This chapter examines a participatory approach to social auditing and codes of labour practice, with a particular focus on the gender dimension. It draws on the experience of Diana Auret as coordinator of the Ethical Trading Initiative (ETI) horticulture pilot in Zimbabwe, and vice chair of the Agricultural Ethics Assurance Association of Zimbabwe (AEAAZ) which developed a National Agricultural Code. During the ETI pilot, the methods and tools of participatory social auditing were developed, which involved the establishment of a local multi-stakeholder initiative in Zimbabwe. The chapter also draws upon a subsequent research project on gender and ethical trade in African horticulture (Smith et al, 2004). The research provided insights into how a participatory approach to social auditing can enhance gender sensitivity and awareness of more marginal insecure workers.[2] The chapter does not cover the practical details of the audit process, as this has been covered in detail elsewhere.[3]

There are a large number of codes of labour practice, implemented by different companies and organizations (see Chapter 1).[4] The better codes of labour practice are based on core International Labour Organization (ILO) conventions covering international labour standards, and require compliance with relevant national legislation. Some codes have been produced by multi-stakeholder initiatives that involve different stakeholder groups, including companies, trade unions and NGOs. Two examples are Social Accountability International (SAI) in the US and the ETI in the UK, both of which are based on core ILO conventions (the key elements of the ETI Base Code are outlined in Chapter 1).[5]

Current social auditing approaches[6]

At a broader level, social auditing is a way of 'measuring and reporting on an organization's social and ethical performance. An organization which takes on an audit makes itself accountable to its stakeholders and commits itself to following the audit's recommendations' (see www.nef.org.uk). A social audit undertaken to ascertain compliance with a code of labour practice is one specific form of audit. It is normally based on measuring and reporting against a buyer's code, or an independent auditable standard. Where this code incorporates core ILO conventions, an audit will be reporting compliance in relation to core international labour standards. A company using the ETI Base Code would employ this as the basis of an auditing checklist, with which all suppliers have to comply.

A social audit can take different forms, and there are normally three types:

1   First party – where a company undertakes an assessment of itself;
2   Second party – where a company audits one of its suppliers against its own code or an external standard;
3   Third party – where a company is audited by an independent and external body.

First-party auditing is often conducted in advance of a visit by an auditor or certification body, although some global companies accept self-assessment by their suppliers against their code. Second- and third-party auditing are most often undertaken either by a global buyer or by a professional auditing company to check compliance by its suppliers.[7] It is often carried out over a fairly short time period, based on a one-off visit, although this will vary according to the requirements of the client and the type of auditing body used. The social auditor will be given or will draw up a checklist for the code that they have been asked to audit against. Sometimes the components of the checklist will be graded according to 'must/should' ('must' referring to international standards and national legislation, while 'should' covers work-related social issues and sector requirements).[8] If a supplier fails an audit on the basis of a non-compliance with a 'must', they are deemed to have failed the audit. A partial non-compliance in a 'must', however, warrants an immediate correction before compliance is conferred. If the supplier fails on a 'should', there is likely to be a similar recommendation to address the issue. Non-compliances or partial compliances are followed by a recommendation for remedial action to be put in place, within an agreed timeframe.

Third-party audits of suppliers against independent codes also provide global buyers with a recognized form of external verification that minimum labour standards are being adhered to within their supply chain. A number of firms now provide social auditing services. These range from not-for-profit organizations set up with the specific aim of auditing amongst workers in developed or developing countries, as well as companies with a long tradition of auditing in management, environmental, and health and safety standards. Financial auditing companies have also moved into corporate social responsibility and social auditing, including Price Waterhouse Cooper and KPMG, although their involvement is said to be diminishing (CCC, 2005).

Limitations of a 'snapshot' approach

The advantage of compliance-based social auditing is that it provides a quick and easy means of assessing the level of compliance. Companies like it because it provides a set of verifiable indicators that can be quantified and checked off in the same way as a health and safety, financial or technical audit. A number of criticisms, however, have been raised regarding narrow compliance approaches. One is that social issues are often difficult to assess for non-compliance. For example, it is possible to observe whether a fire extinguisher is accessible in any emergency, but it is not possible to determine if a worker has suffered discrimination or abuse by a supervisor. Some professional auditing companies use foreign auditors who pay brief 'flying' visits, which diminish their ability to genuinely assess the types of employment grievances that might exist. Furthermore, standard approaches that are company compliance based are carried out from the perspective of management, and often only superficially seek the views of workers in relation to their employment conditions. Finally, the supplier normally (though not always) pays for the audit, and this can put a severe strain on small companies and/or those that supply multiple buyers and must pay for repeated audits against different codes.[9]

Audits are normally confidential; but one open study was undertaken of two Price Waterhouse Cooper (PwC) audits conducted in factories in China and Korea.[10] This study provided a detailed critique of the monitoring systems and methods used by PwC auditors. It argued that PwC auditors relied on information gathered primarily from managers rather than workers, and depended largely upon data provided by management with little or no 'triangulation' or cross-checking. The study also found that the auditors failed to pick up a number of issues relating to freedom of association, health and safety, wages and hours. These types of revelations have placed snapshot social auditing increasingly under criticism for failing to adequately address workers' rights (CCC, 2005).[11] As noted, one dimension of the problem is that audits are too management focused. Many conventional auditors spend long periods with managers, allow managers to select workers, fail to explain the purpose of their visit to workers, and only hold brief worker interviews. In this context, workers do not feel at ease to talk openly to auditors and may even feel intimidated. The other dimension of the problem is that where audits are a regular occurrence, some firms have become increasingly adept at avoiding auditor detection of issues. This can be done by telling some workers, such as casual and third-party contract workers, not to come in on audit days; informing workers that they may lose their contracts if they do not give a good impression; coaching workers to give auditors the 'correct' answers;[12] and only selecting favoured workers for interviews by auditors. In some countries, firms commonly go further and hold double sets of books in order to hoodwink auditors on issues such as pay and overtime (CCC, 2005).[13] In a competitive commercial environment where producers are under pressure from retailers to meet strict deadlines for order delivery, it can be difficult to meet social compliance and the standards set by codes (Oxfam, 2004). As a result, many firms have adopted a management-based approach to convince buyers that they are compliant with standards, but have little obligation to the workers' rights contained in codes of labour practice.

Developing a participatory approach to social auditing

Participatory social auditing (PSA) has been developed as a means to address these problems (Bendell, 2001; Auret, 2002). In contrast to a compliance-based approach, PSA views codes as one means of improving working conditions and enhancing workers' rights. PSA is also based on a different philosophy, one that is worker centred and process oriented, and that aims to instil learning and improvement rather than simply checking for one-off management compliance.

Box 8.1 Motivation for use of Participatory Social Auditing

Participatory Social Auditing (PSA):

- encourages the active involvement of workers and managers in the auditing process;
- enables discussion, allowing the freedom for both workers and employers to share their ideas and perceptions about labour and social issues;
- creates awareness and enables people to identify problems and priorities;
- encourages the building of better relationships between workers and employers;
- enables joint planning and decision-making;
- enables access to a considerable number of people in a day.

Source: Auret, 2002

A participatory approach adapts and applies tools from Participatory Rural Appraisal (PRA) and Participatory Learning and Action (PLA), developed in the disciplines of anthropology and rural sociology to the process of social auditing.[14] It emphasizes the importance of using local auditors trained in the use of participatory tools, who understand the language, culture and local issues, and who understand the ways in which workers can be exploited. A participatory approach also emphasizes the importance of engaging local stakeholders, especially trade unions and NGOs, outside the company. This helps to check or triangulate information, and to ensure that all issues concerning workers are being raised. This approach is particularly important in relation to sensitive issues such as freedom of association, discrimination, verbal abuse, sexual harassment and child labour – issues that are often difficult to discern from documentation or through formal questioning.

A gender-sensitive approach to social auditing

In many export sectors women form the majority of insecure workers (seasonal, contract or casual) and face heightened employment vulnerability. Particular problems include unequal wages, lack of freedom of association, discrimination in training and promotion, verbal and sexual abuse by supervisors, and difficulties in combining childcare with long hours and compulsory overtime (often at short notice).[15] Underlying the challenge of identifying these sorts of problems faced by women in insecure and vulnerable employment is their lack of 'empowerment' as workers.[16] This can be reflected at different levels. At an institutional level, women in insecure work often lack protections in labour law that would allow them to legally claim rights as workers. At an employment level, they often lack access to information that would enlighten them to their rights, either by law and/or through a code of labour practice. Women's lack of organization or representation reinforces this position and ensures that they lack the ability to influence change. At a personal level, women's employment conditions are often embedded in gendered norms and practices which they have been conditioned to accept; therefore, they are less likely to challenge a situation, even where it is in violation of the law and/or a code of labour practice.

As a consequence, women workers, particularly where their work is insecure, often lack awareness of their rights and the confidence to voice complaints to outsiders. In addition, few 'gender issues' are explicitly covered in codes of labour practice, and many gender issues are of a sensitive nature. It is therefore necessary for an auditor to be thoroughly aware of such issues, and to be adequately trained in the use of appropriate methods and tools in order to access such information. In addition, past experience has shown that traditional audit tools do not succeed in uncovering many sensitive gender issues. To do so requires new participatory tools that can encourage reluctant workers to reveal their own issues.

> ### Box 8.2 Key issues for a gender-sensitive approach to social auditing
>
> The following characteristics, typical of female workers in developing coun-
> tries, are of key importance when considering the nature of the approach and
> methods to be used in the social audit process if the 'developmental' aspect is
> to be realized:
>
> - low literacy level;
> - lack of awareness of their rights as workers in civil society;
> - cultural norms and beliefs that dictate the subordinate role of women in
>   society.
>
> *Source:* Auret and Barrientos, 2004

Participatory methodology, with its wide variety of techniques and tools, was used in the ETI pilot in Zimbabwe and was found to be very successful at gathering data from a wide range of management and workers, especially women. Female workers gained confidence from the participative approach, which encouraged them to 'open up' (particularly when working in groups), resulting in a number of key gender issues being highlighted. One of these issues was 'sexual harassment', which had not been identified in the previous 'snapshot' social audits conducted in Zimbabwe. This shows that a participatory approach can detect less visible and more sensitive issues of non-compliance, which a snapshot approach is less able to pick up.

Gender-sensitive social auditing as a process

Central to a participatory approach is the process involved, of which the final social audit is an outcome rather than the means in itself. This process involves various stages. The first and most important is that of awareness creation amongst employers and workers. The second is the pre-audit, where issues are revealed and assessed, accompanied by engagement with the employer, workers and worker representatives to develop an implementation plan that will lead to improvement. The third stage is the final audit, where the employer is formally assessed for compliance. If the first and second stages have been effective, passing the audit should be the logical final outcome. The philosophy from the beginning is helping the employer to enhance employment conditions and address workers' rights, rather than being policed or reprimanded for failure to comply.

The social audit process focuses on the involvement of both management and workers in each and every phase of the process. Such involvement serves to heighten the consciousness of the key role played by the workers, male and female, and of the interdependence of management and workforce in the audit process. Given the largely female workforce in global export industries, it is necessary to constantly adapt the process to ensure adequate female worker participation. This must take place at each step of the social audit. In this section we will examine the three stages of a gender-sensitive participatory social audit: awareness creation, pre-audit and final audit. In 'Participation and a multi-stakeholder process', we will examine the specific participatory tools that can be used in the process.

Awareness creation

Creating an awareness of the standards and principles of codes among local producers and workers is important for increasing their familiarity with a code of labour practice. In addition, it leads to:

* heightened awareness of related worker rights and legislation;
* increased understanding that producer and worker are both involved in the process;
* a growth of commitment to comply with the standards and the development of a sense of ownership of the code.

Creating awareness of code principles also raises awareness of other issues, such as the importance of communication between management and workers. Few producers or senior staff are aware of the 'managerial significance' of communicating with their workforce, particularly when it is predominantly female. This relates particularly to policy- and work-related information that is normally passed through a 'chain of command' involving middle and junior managerial staff who are predominantly male. This chain of command often creates barriers to communication from senior management to workers, who are often insecure, female workers. For this reason, normal communication channels cannot be relied upon to create awareness with regard to the principles of the code of labour practice, and other means need to be found.

Pre-audit

The pre-audit provides an opportunity, especially for workers who would have had little or no previous involvement in audits, to become familiar with the various steps in the process, their role in aspects of the final audit and the participatory tools and techniques to be used. The latter is of considerable importance given that the objective is to maximize the involvement of workers, particularly women and those in insecure employment. Participatory tools drawn from PRA and PLA are used in order to encourage workers to participate and share information. Such tools include focus group discussions (FGDs) of eight to ten workers, with both female, male and mixed groups, where workers are encouraged by a facilitator to discuss issues between themselves. Within FGDs, participatory tools such as mapping, ranking and scoring exercises are used to encourage interaction between workers in representing issues as they see them. Box 8.3 provides a summary of participatory tools.[17] Participatory tools not only offer an opportunity for informal sharing of information, but also build confidence in female workers in their capacity to successfully participate in the auditing process – of crucial importance for a successful final audit.

The timing of the audit is also important in ensuring that temporary, casual, migrant and other non-permanent workers are available for interviews. This means that audits usually have to be undertaken at the peak of the season or of production activity. While this might create resistance from management, who would prefer a slacker period, there is a risk that only permanent workers are on site during the off season, and that the audit would miss more vulnerable workers who are often women. The records check at the end of the pre-audit also provides an opportunity for the auditor to investigate key gender issues, such as 'equal pay for equal work' or 'equal access to training and promotion', as well

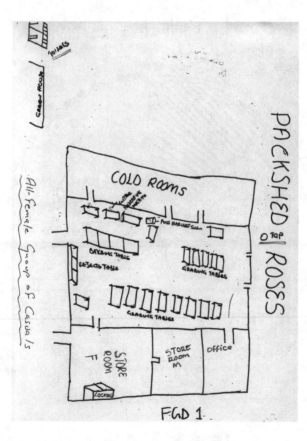

Notes: Figures 8.1 and 8.2 provide examples of a mapping and wheel diagram as drawn by participants in two FGDs during research on ethical trade in African horticulture.

Figure 8.1 *Mapping*

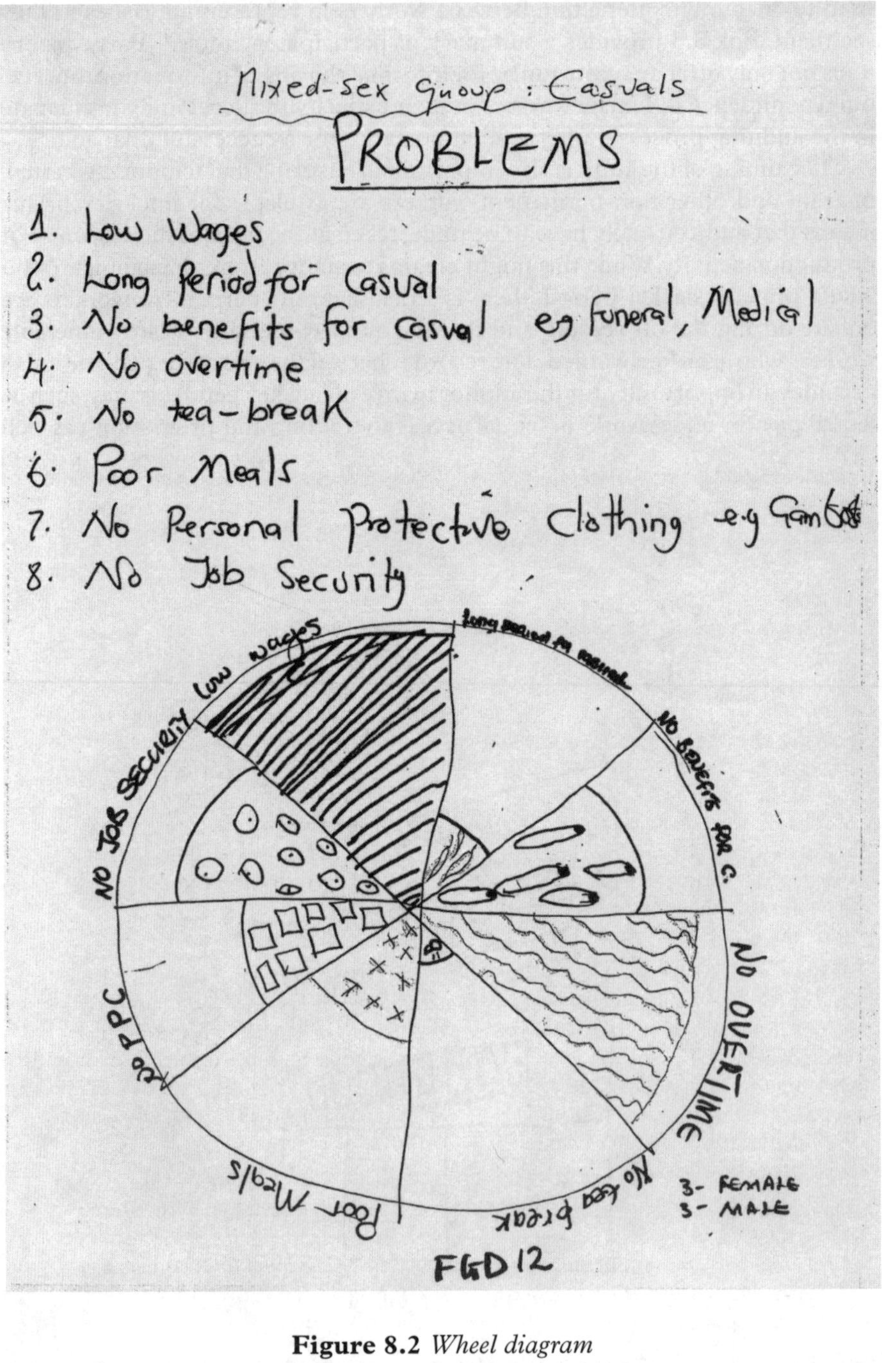

Mixed-sex group : Casvals

PROBLEMS

1. Low Wages
2. Long Period for Casual
3. No benefits for casval eg funeral, Medical
4. No overtime
5. No tea-break
6. Poor Meals
7. No Personal Protective Clothing e.g Gumboots
8. No Job Security

3- FEMALE
3- MALE

FGD 12

Figure 8.2 *Wheel diagram*

Box 8.3 Examples of participatory tools used in social auditing

Participatory tool

Gender responses

Mapping
Visual representation or diagram of a work area.
Identifies issues of importance.
Illustrates differences in perceptions by different groups.

This helps to depict the layout of a specific work area or the worksite as a whole: men are often better at the latter than women, as spatial perceptions are linked to literacy and exposure to 'a larger world'. However, once women understand what is required they can very easily draw their own working area.

Role play
Workers enact experiences important to them from their working lives.
Helps a group of workers to express their realities and problems.

Women often enjoy role play – especially in relation to any sensitive issue, e.g. sexual harassment, job harassment from male supervisors, late working hours or lack of money to care for their children.

Wheel
Issues for discussion scored/ranked by group.
Wheel drawn and divided into equal-sized labelled segments.
Scoring of each segment is discussed by the group.
Facilitates observation and discussion of what is important, why it is important, how important it is, and which groups are affected.
Provides documented information from the group discussion.

Both men and women can draw the Wheel well. Women appear to like the Wheel because (a) its roundness is like their perception of life itself; and (b) they like to participate in the drawings within each segment of the wheel. If men use a wheel for ranking issues they often fill in the segments using lines and angular designs, while women will draw patterns, ranging from flowers to children, to circles and birds.

Spider
Spider drawn by a group participant.
Spider legs and antennae labelled with key issues for workers.
Encourages discussion of key issues of concern to workers.
Provides documented information from group discussion.

Men prefer using the spider to rank as it can be done by one person, although in most cases the group members are all encouraged to 'draw a leg'. If given a choice, women who do not like spiders in real life would rather not use it.

Transect walk
Walk across a work area or farm with a worker/workers.
Observe and discuss what you see with the person(s) you walk with or those whom you meet.
Often adds oral data to information mapped.
Highlights worker experience of issues.

Generally undertaken with two or more participants from the focus group discussions (FGDs) because, once again, it's natural to walk and talk in a group.

Source: Auret and Barrientos, 2004

as to verify data obtained verbally or visually. It can help to reveal the degree of female workers' involvement in trade unions and/or workers' committees, where they can participate in meaningful decision-making processes.

Feedback and remediation

Management feedback of the results of the pre-audit to workers (male and female) is crucial to the success of the final audit. Feedback should include both positive findings and those areas that require improvement, as well as the remedial action to flow from the pre-audit. Feedback promotes worker trust in the auditors since it is at this stage that they learn that their anonymity has been kept – especially important to vulnerable workers who may be afraid to lose their jobs. It also facilitates management discussion of ways and means 'to improve' the existing situation. This indicates to the workers that their observations and perceptions have been taken seriously by management, and that female workers are important to the company. Ensuring that non-permanent members of the female workforce are adequately represented in the feedback process requires careful planning.

Final audit[18]

Prior to the final audit report, it is recommended that a general meeting be held with management and worker representatives (including women workers) to inform both management and employees of its details. The final audit is generally undertaken approximately six months after the pre-audit, but can be at any time within the following year. It goes through the same stages as the pre-audit, other than including feedback and remedial action recommendations.

Composition of the social audit team

The audit should be carried out by two independent, appropriately qualified and experienced social auditors; this allows one auditor to focus solely on the facilitation of focus group discussions, while the second records all data elicited. The use of two, rather than one, social auditors is particularly necessary if the participatory methodology is used. Experience in the ETI pilots and research project highlighted the importance of using local social auditors who are:

- able to speak the language of the interviewees;
- aware of the cultural background of the workers concerned;
- knowledgeable about the country legislation and constitution;
- qualified and experienced in the use of the participatory methodology.

The gender composition of the workforce has particular significance for whether the social audit team should be male and/or female. Wherever there are large numbers of female employees, the social audit team should be composed of two female auditors in order to:

- facilitate ease of access to female interviewees;
- create a conducive, accepting environment for the sharing of sensitive issues;

- encourage the participation of female employees, especially the shy and semi-literate.

How the auditors approach and relate to workers, especially those who are more vulnerable and, possibly, with low literacy and confidence levels, is also very important. A formal, authoritative style of interviewing can easily lead workers into remaining silent and disengaged from a social audit. The use of participatory tools when conducting worker interviews, however, can help workers to open up and engage more actively in the process. Finding local auditors with the relevant experience and characteristics can be a challenge, especially for an individual company. This is particularly so in countries or sectors where participatory social auditing has not been used before. Participatory methodology is increasingly used in social science and civil society research in many countries, creating a network of people with relevant understanding of the tools, but not necessarily social auditing per se. Alternatively, local social auditors may be available who have not been trained in the use of participatory tools. In the initial stages, selection and training of local participatory social auditors may be a requisite step. This could be done by individual companies, but would require resources. Another approach, which will be examined in the 'Conclusions' section, is the pooling of resources through the development of local multi-stakeholder initiatives able to develop this capacity.

Participation and a multi-stakeholder process

The use of participatory tools in the audit process can play an important role in highlighting gender issues, which a more formal snapshot audit can fail to uncover. But the use of such tools is only one aspect of a wider participatory approach to codes of labour practice. The philosophy behind this approach is that workers should not simply be passive objects of an external audit, but should become more actively engaged in a process of improvement of their working conditions. Worker engagement can be extended through developing ongoing local independent monitoring and verification involving worker representatives as part of a more sustainable approach.

An important limit to worker engagement in social auditing is their insecure economic position relative to their employer or senior management. Workers individually are often fearful of losing their jobs if they express their concerns too openly. This is particularly so for insecure workers, who are more likely to experience labour abuse. Social audits must involve, wherever possible, trade unions. Shop stewards and elected worker representatives should be interviewed as part of the pre-audit process, in awareness creation and during the audit. They should also be involved in feedback and remediation at every stage. They are able to play a vital part in ensuring that compliance is maintained, and that improvements made in working conditions are sustained after an audit is completed.

At a broader level, a participatory approach to social auditing should include the involvement of industry trade unions and other relevant NGOs. This is particularly important where an individual site has little or no unionization. It can be done through consultation prior to an audit in order to identify the key issues facing workers in a particular sector, including important gender issues. Auditors are then alert to potential problems of non-compliance when they visit a site. However, local trade unions and NGOs often lack the resources or capacity to become involved in numerous individual social audits in a large number of companies, given the many other pressures on their time.

Social auditing that forms part of a 'process', rather than a one-off visit by auditors completing a checklist inspection, can be further extended through the development of local multi-stakeholder initiatives.[19] Local country codes of labour practice, promoted and supported by a multi-stakeholder association that includes industry, trade unions, relevant NGOs and government representatives, would provide greater local accountability. Multi-stakeholder initiatives extend beyond the concept of compliance with international standards by individual firms, and focus more widely on the sustained improvement of the working and living conditions of workers of a particular sector or country. Local multi-stakeholder initiatives (see Box 8.4) are able to draw on the wide range of knowledge and experience of the different stakeholders in a sector or country. They involve participants who have in-depth knowledge of relevant national law, industry, employment and social context, as well as key issues facing workers. They can play a role both in relation to facilitation, and monitoring and verification of a code of labour practice. The services of such a body can help in:

- the promotion and implementation of codes of labour practice;
- the development of a harmonized code that is locally relevant;
- the organization and provision of services such as awareness creation, pre-audits and remedial action advice to producers;
- the provision of information services to external stakeholders, industry associations and other related bodies, nationally and internationally;
- the task of contracting local independent, qualified social auditors to undertake social audits;
- ongoing independent monitoring that involves worker representation.

All of the above contribute towards the improvement of labour conditions and standards, of management–worker relations and, ultimately, the growth of more ethical trading. The development of local multi-stakeholder initiatives is still at a fairly early stage and, as Box 8.4 indicates, there are many challenges involved.

Advantages and challenges of a local multi-stakeholder approach

Developing local multi-stakeholder initiatives has potential benefits over a more formal externally audited compliance-based approach. The setting up of an independent body that represents different stakeholders provides a degree of

Box 8.4 Examples of local, multi-stakeholder initiatives

Several multi-stakeholder initiatives have merged to guide social accountability in the African horticulture industry. These include the following.

Agricultural Ethics Assurance Association of Zimbabwe (AEAAZ)

The Agricultural Ethics Assurance Association of Zimbabwe (AEAAZ) was set up following the Ethical Trading Initiative (ETI) pilot project in Zimbabwe. It is an autonomous body, governed by representatives of producers, trade unions and non-governmental organizations (NGOs), which aims to promote and ensure compliance with the Zimbabwean National Agricultural Code of Practice. It seeks to improve social, chemical and environmental standards on agricultural export farms, with a view to maintaining and improving access to export markets. By the end of 2001, 300 producers had registered with the association to implement the code and be certified to export to Europe and the UK. The success of the ongoing process of awareness creation, pre-audits and final audits using the same participatory methods resulted initially in a rapid expansion of membership, due in large measure to the vastly improved relationship between employer and employee, and the increasing number of growers who complied with the social code. This was reinforced by overseas buyers' acceptance of the AEAAZ social accreditation. Unfortunately, in 2002, the political problems in Zimbabwe, and particularly the land invasions in agriculture, subsequently restricted its work (Auret, 2002).

Wine Industry Ethical Trade Association (WIETA), South Africa

A similar type of body was officially launched in 2003 in South Africa, following the ETI wine pilot in the Western Cape. The Wine Industry Ethical Trade Association (WIETA) is an autonomous body which all companies, trade unions and NGOs linked to the wine industry, and committed to its objectives, can join. It has an executive committee that comprises representatives from each constituency group, as well as government, and has its own code of practice (linked to the ETI base code and South African legislation) and sets out in more detail what each of the ETI principles entails within the wine industry context. In addition to formulating and promoting the adoption of a code of good practice among wine producers, WIETA has developed social auditing and monitoring procedures for its member wine farms and estates, and may extend to include fruit farms.[20] By 2005, WIETA membership had grown to 85 members, including wine distilleries, vineyards, trade unions, NGOs and government. It was extended to include all sections within agriculture. It has been recognized by a number of UK supermarkets, four of whom are now also members of WIETA.

Horticultural Ethical Business Initiative (HEBI), Kenya

During 2002, local civil society organizations spearheaded a campaign against poor working conditions on Kenyan flower farms, generating concern about the reputation of the industry in overseas markets. As a consequence of this activity, a steering committee – the Horticultural Ethical Business Initiative (HEBI) – was formed to guide social accountability in the horticulture sector. HEBI initiated a multi-stakeholder approach to code implementation and formed a tri-partite Stakeholders Steering Committee (SSC) comprised members from civil society organizations and trade associations/employers, four observers, and government representatives as regulators. The SSC was given two broad terms of reference: first, to harmonize stakeholder interests and involvement, and to develop a participatory social audit system acceptable to all stakeholders, including overseas buyers; and, second, to use the social audit system to assess the social conditions on flower farms and to establish a baseline for future activities. In 2003, following the development of the draft HEBI code, HEBI applied Participatory Social Auditing (PSA) methods in eight pilot social audits (with a multi-stakeholder audit team) as part of a learning initiative to trial participatory methods in social auditing (Dolan and Opondo, 2005).

independence from any one organization dominating the process. Importantly, it facilitates workers' representation through trade union participation, giving voice to workers who may normally be fearful of speaking out. The involvement of NGOs can help to ensure that sensitive issues, such as gender and racial discrimination, are raised, as well as work-related issues such as childcare and social provision. Where government is involved, it can also provide a bridge between voluntary initiatives based on codes, and national regulation and enforcement of labour standards. Such a body can provide guidance, oversee code implementation and facilitate auditing on a basis that is mutually agreed between all stakeholders. Bringing different stakeholders together is important in building new alliances and forging new ways forward.

Developing local multi-stakeholder initiatives also poses significant challenges. One of the main obstacles to developing a local initiative is resourcing, particularly during the early phase when it does not have sufficient membership to generate funds of its own. Which stakeholders should legitimately be involved, and how the process is managed, will vary in different locations, with no set formula for success. In some countries not all trade unions are independent of employers or the state with democratically elected officials. NGOs are a heterogeneous grouping with no formal accountability to workers, and some NGOs do not have the same level of local community engagement as others. There are issues around the relative power of different groups and the economic resources available to employers relative to trade unions and

NGOs. There are often inherent tensions between stakeholders (as there should be if they are reflecting their constituency interests), with deeply held suspicions on different sides. These can be difficult to overcome, requiring careful management, and it often takes time to build trust.

From a gender perspective, local multi-stakeholder initiatives also present both opportunities and challenges. Women in insecure employment are often the least likely to be organized or unionized, reinforcing their vulnerability. Multi-stakeholder initiatives that involve trade unions and NGOs sensitive to the needs of such women workers are more likely to ensure that the gender issues facing such workers are addressed. The combination of participatory social auditing and independent monitoring based on stakeholder engagement can help to give voice to such vulnerable workers. However, gender discrimination is often deeply embedded in employment practice and social relations. Local organizations (trade unions and NGOs) that operate in the same social context can also serve to reinforce existing gender norms. Hence, while local multi-stakeholder initiatives can open up the space for gender and racial discrimination to be addressed, this is not automatic. They create an opportunity, but not a guarantee, for the enhanced participation of insecure women workers in the implementation of codes.

Developing a harmonized local code can also help to overcome the problem of proliferation of diverse codes that face many suppliers. Engagement by trade unions and NGOs sensitive to gender issues can help to ensure that a harmonized code addresses core labour standards and gender discrimination. But there remains the challenge of international recognition by buyers, even if the code is benchmarked against an internationally recognized code such as that of the ETI and incorporates all core ILO conventions. All of the local initiatives in African horticulture discussed in Box 8.4 have faced this difficulty, and it has not been easy to overcome (even though their linkages to the ETI have helped in relation to the UK market). Despite these challenges, local multi-stakeholder initiatives are becoming more established and represent an important move away from a Northern-led, top-down approach to codes of labour practice.

Conclusions

A participatory approach to social auditing is viewed as part of a process that involves awareness creation and dialogue between employers, workers and their representatives. It starts from the principle of improving working conditions and enhancing workers' rights, rather than focusing on minimal management compliance. It aims to ensure that more insecure and vulnerable workers, who often have low confidence and literacy levels, have a voice in social audits. It involves the use of tools drawn from Participatory Rural Appraisal and Participatory Learning for Action. It stresses the need to use local auditors, with local knowledge and language, who are sensitive to gender issues. This is in contrast to the culture of 'snapshot' social auditing, based on brief formal visits by outside professional auditors. Here the focus is on 'policing' suppliers, who, in turn, carry out the minimum changes needed to pass an audit, rather than making sustainable improvements in working conditions. Through awareness creation, a participatory approach attempts to change the 'mindsets' of employers and increase workers' understanding of their rights. The focus is not only to ensure that minimum labour standards are met, but to ensure that improvements in employment practices reach all groups of workers. Such an approach faces many challenges; but it represents a shift away from a formal top-down compliance orientation to the greater empowerment of workers and their representative organizations as an essential part of the process of improving labour standards and working conditions.

Notes

1   The research underpinning this chapter was funded by the UK Department for International Development (DFID) SCR. DFID supports policies, programmes and projects to promote international development. DFID provided funds for this study as part of that objective; but the views and opinions expressed are those of the authors alone.

2   One important aspect of Smith et al's (2004) study was to explore the manner in which women and men responded to different research tools used in social auditing. It examined, for example, the different interactions between same-sex or different-sex auditors and participants in individual interviews and group discussions. The study also identified a range of issues that had important gender implications, both for the preparatory stages of an audit and for the methods and tools used in the actual audit.

3   This chapter is drawn from a longer Institute of Development Studies (IDS) *Working Paper* (Auret and Barrientos, 2004), which also includes more practical information on how to conduct a participatory social audit. For more details on participatory social audits, see this paper and Auret (2002). These are also complemented by a training video *Workers Matter: A Participatory Approach to Social Auditing*, available through www.ids.ac.uk/ids/bookshop.

4   For more detailed discussion of codes of labour practice, see Jenkins et al (2002); Urminsky (undated) and Utting (2002).

5   See www.ethicaltrade.org for more details.

6   This section focuses specifically on social auditing of codes of labour practice applying to global suppliers, and not the wider practice of social auditing and accounting of organizations and firms (see www.caledonia.org.uk for more general references on social auditing). The section draws on Barrientos (2003).

7   In this chapter we are focused on the methods used by external auditors engaged in second- or third-party audits.

8   There is no set standard for social auditing, and different companies use different criteria; this reflects one commonly used approach.

9   One Chinese firm reported being audited by teams from 40 customers in a single month, from a combination of buying firms, external audit firms and NGOs; see Kaplinsky and Morris (2001).

10   A detailed critique of professional social audits carried out by Price Waterhouse Cooper of garments factories in Shanghai and Seoul is given by O'Rourke (2002).

11   One example was a BBC 'Panorama' programme (15 October 2000), which exposed breaches in the codes operated by Gap and Nike in a factory in Cambodia relating to hours of work, overtime, wages and child labour. The factory had been audited, but it had failed to identify these problems, which a Panorama team did through outside interviews with workers.

12   When workers start to answer an auditor's questions before they are completed, this is a sure indication that workers have been coached (pers comm with an auditor in China, confidential source).

13   It is reported that in China there is now a software programme available that can keep up to 16 sets of 'books' to comply with different codes on hours, pay and related issues, with errors built in to deter auditors from suspecting that they have been falsified (pers comm with an auditor in China, confidential source).

14   For more information on Participatory Rural Appraisal, see Chambers (1997). Further resources on PRA and PLA can be found on www.ids.ac.uk/ids/particip/ information/index.html and www.iied.org/resource.

15   For an overview of female employment levels and issues, see Barrientos et al (2003). For a more detailed discussion of gender issues relating to codes of labour practice, see Smith et al (2004).

16   The term 'empowerment' is used in different ways. Here it is used to relate to the choices and actions people can make on their own behalf from a position of economic, political and social strength. For further discussion, see Kabeer (1994); Rowlands (1997); VeneKlasen and Miller (2002).

17   For more detailed examination of participatory tools that can be used in social auditing, see Auret (2002) and Auret and Barrientos (2004).

18   A more detailed outline of the methods for undertaking a final audit is contained in Auret (2002).

19   The term 'stakeholder' is used here to refer to 'any individual, community or organization that affects or is affected by the operations of a company... All stakeholders are not equal and should not be treated equally. The workers whose working conditions are the subject of codes of labour practice are recognized as having the greatest "stake" in ethical trading' (ETI, 2003, p107).

20   For more detail, see the WIETA webpage on www.wosa.co.za/wieta.asp. For background information on the ETI Wine Pilot project, see ETI (2004).

References

Auret, D. (2002) *Participatory Social Auditing of Labour Standards: A Handbook for Code of Practice Implementers*, Agricultural Ethics Assurance Association of Zimbabwe, Harare

Auret, D. and Barrientos, S. (2004) 'Participatory social auditing: A practical guide to developing a gender-sensitive approach', *IDS Working Paper*, no 237, Institute of Development Studies, Brighton

Barrientos, S. (2003) *Corporate Social Responsibility, Employment and Global Sourcing by Multinational Enterprises*, paper for International Labour Office (ILO), Geneva

Barrientos, S., Hossain, N. and Kabeer, N. (2003) 'The gender dimensions of the globalisation of production', Background paper to the World Commission on the Social Dimensions of Globalization, International Labour Office (ILO), Geneva

Bendell, J. (2001) *Towards Participatory Workplace Appraisal: Report from a Focus Group of Women Banana Workers*, New Academy of Business, Bristol

CCC (Clean Clothes Campaign) (2005) *Looking for a Quick Fix: How Weak Social Auditing Is Keeping Workers in Sweatshops*, CCC, Amsterdam

Chambers, R. (1997) *Whose Reality Counts? Putting the First Last*, Intermediate Technology Publications, London

Dolan, C. and Opondo, M. (2005) 'Seeking common ground: Multistakeholder initiatives in Kenya's cut flower industry', *Journal of Corporate Citizenship, Special Issue on Corporate Citizenship in Africa: Progress Looking Back; Prospects Looking Forward*, vol 8, summer, pp87–98

ETI (Ethical Trading Initiative) (1998) *Purposes and Principles*, ETI, London, www.ethicaltrade.org

ETI (2003) *ETI Workbook: Step by Step Guide to Ethical Trade*, ETI, London

ETI (2004) *Inspecting Labour Practice in the Wine Industry in the Western Cape, South Africa, 1998–2001*, ETI, London, www.ethicaltrade.org/Z/lib/2004/02/sawine-rept/index.shtml, accessed August 2004

ILO (International Labour Organization) (2002) *Decent Work and the Informal Economy, International Labour Conference*, 90th Session, ILO, Geneva

Jenkins, R., Pearson, R. and Seyfang, G. (eds) (2002) *Corporate Responsibility and Labour Rights: Codes of Conduct in the Global Economy*, Earthscan, London

Kabeer, N. (1994) *Reversed Realities*, Verso, London

Kaplinsky, R. and Morris, M. (2001) *A Manual for Value Chain Research*, International Development Research Centre (IDRC) and Institute of Development Studies, Sussex, www.ids.ac.uk/ids/global/ man&hand.html, accessed 6 December 2004

O'Rourke, D. (2002) 'Monitoring the monitors: A critique of third-party labour monitoring', in Jenkins, R., Pearson, R. and Seyfang, G. (eds) *Corporate Responsibility and Labour Rights: Codes of Conduct in the Global Economy*, Earthscan, London, pp196–208

Oxfam (2004) *Trading Away Our Rights: Women Working in Global Supply Chains*, Oxfam, Oxford

Rowlands, J. (1997) *Questioning Empowerment: Working with Women in Honduras*, Oxfam, Oxford

Smith, S., Auret, D., Barrientos, S., Dolan, C., Kleinbooi, K., Njobvu, C., Opondo, M. and Tallontire, A. (2004) 'Ethical trade in African horticulture: Gender, rights and participation', *IDS Working Paper*, no 223, Institute of Development Studies, Brighton

Urminsky, M. (ed) (undated) *Self Regulation in the Workplace: Codes of Conduct, Social Labelling and Socially Responsible Investment*, International Labour Office (ILO), Geneva

Utting, P. (2002) 'Regulating business via multistakeholder initiatives: A preliminary assessment', in *Voluntary Approaches to Corporate Responsibility: Readings and a Resource Guide*, United Nations Research Institute for Social Development (UNRISD), Geneva

VeneKlasen, L. and Miller, V. (2002) *A New Weave of Power, People and Politics*, World Neighbours, Oklahoma, www.wn.org

9

Oxfam's Coffee Campaign: An NGO Perspective

Henk Campher

Introduction

The coffee sector has experienced a major crisis over the past decade. The dismantling of the international coffee agreement in 1989, coupled with market liberalization, corporate consolidation and global oversupply, plunged coffee commodity prices to their lowest levels in a century (Ponte, 2002). The price of coffee fell by almost 50 per cent between 1999 and 2002. The crisis affected the livelihoods of 25 million coffee producers around the world, mostly poor smallholders, who sell their coffee beans for much less than they cost to produce. In September 2002, Oxfam launched a worldwide campaign to draw attention to the plight of coffee farmers. As part of its international Make Trade Fair campaign, Oxfam called for the major players in the coffee industry to support a Global Coffee Rescue Programme and create a more just system of trade for small producers. This chapter examines the problems facing coffee producers in developing countries, the role of large multinationals operating within the coffee sector, and explains Oxfam's campaign to develop a Coffee Rescue Plan that aims to support coffee producers and workers.

Context of the coffee campaign

The Oxfam International coffee campaign aimed to highlight the plight of the 25 million coffee farmers whose livelihoods were being destroyed by the collapse of the global price of coffee (Oxfam, 2002). Oxfam's Coffee Rescue Plan was intended to meet the farmers' short-term need to endure the crisis and to address longer-term solutions to the market's overproduction. Soon after, the Global Alliance on Coffee and Commodities (GLACC) was launched, bringing together about 100 producers, consumer groups, trade unions and environmentalists. GLACC joined Oxfam in calling on major players in the coffee market – many of whom were making comfortable profits from the coffee business – to help reform the failing market.

Initial responses to the call for reform were promising. Governments across the globe – Spain, Belgium, The Netherlands, Germany, Brazil, the UK, Colombia, Peru and Ethiopia – welcomed the campaign and sent messages of support and commitment. The UK government initiated a debate on commodities, as did the European Union, and the European Commission developed an action plan for tackling the crisis in commodities. In the US, both the Senate and the House of Representatives ratified formal declarations on the urgent need to address the coffee crisis, and they rejoined the International Coffee Organization (ICO) in 2004. The German government, backed by many in the industry, developed a Common Code of Conduct for Coffee (CCCC) for the sustainable production of coffee. African, Caribbean and the Pacific countries, for their part, have attempted to persuade the European Union to use 750 million Euros of unspent aid money to help farmers either move out of the coffee market or strengthen their position in it (see Oxfam, 2002).

Since the campaign's launch, however, the actions of key stakeholders have been *ad hoc* at best and, in many cases, farmers' poverty has worsened. With its farmers' livelihoods heavily reliant on coffee production, seasonal employment in Central America dropped by more than 21 per cent and permanent employment declined by 54 per cent between 1999 and 2002. Since February 2003, more than 5000 Guatemalan families have relied on food aid to avoid hunger. In El Salvador, debts totalling US$340 million owed to state institutions by 23,000 coffee growers have proved so untenable that in 2003 the government announced a two-year freeze on payments of both interest and capital. The problem is not confined to Latin America. In African countries, many farmers are finding it desperately difficult to compete in low-priced markets and are simply withdrawing from coffee production. These farmers have few alternative means of survival. The countries hardest hit – Cameroon, Ivory Coast and Tanzania – all qualify as least-developed countries and, consequently, suffer from high levels of infant and child mortality, limited access to healthcare and education, and endemic poverty. In these countries, every dollar lost in coffee production has a multiplier effect, as investments and demand also fall. The losses are felt both in villages, where lower prices mean less food and less money for medicines and education, and at a national level, where lower coffee revenues lead to fewer exports and more debt.

The plight facing today's coffee farmers is a direct result of depressed prices. The composite price indicator (a global calculation by the ICO) suggests that coffee sold at less than US$0.60 per pound in January 2004. Despite differences in the cost of production across countries, this figure means that few coffee farmers, including those with more industrialized production methods, are breaking even. Furthermore, poor farmers are not moving out of coffee production because they have few, if any, alternatives as Northern protectionism and limited infrastructure have dramatically limited opportunities for diversification. Small farmers in developing countries therefore confront a problem of trade and a problem of development. While the coffee companies (both roasters and traders), producer and consumer governments, coffee retailers, multilateral institutions and coffee producers' groups all agree that there is a structural problem in the coffee market – too much coffee is being produced – the initial response was slow, a situation Oxfam has attempted to redress.

The restructuring

Between 1999 and 2003, prices for 'green' un-roasted coffee reached their lowest point in 100 years.[1] While the internationally quoted price for Arabica coffee on the New York Coffee, Sugar and Cocoa Exchange briefly dipped below US$1 per pound during the 1980s, by the turn of the century prices had sunk to unprecedented depths. At its lowest point in 2001, the coffee price fell below 42 cents a pound, and in January 2004 it was still averaging less than 60 cents (ICO, www.ico.org/prices/p2.htm). This decline in prices, in part, results from the chronic instability of commodities prices, which have plagued farmers for centuries. As an agricultural crop, coffee is vulnerable to the weather, particularly changes in temperature and rainfall, as well as disease, making its supply unpredictable. Furthermore, coffee is very slow to respond to price fluctuations. A coffee tree requires five to six years to reach a full yield, and then has a productive life of 15 to 20 years. When prices are high, it takes a long time before production can increase to meet demand, thus strengthening the upward pressure on prices. At the same time, the new production often coincides with a downturn in the market, exaggerating the fall in prices. Moreover, farmers must plant and tend trees for several years before they generate an income. Therefore, when prices fall, farmers are reluctant to cut production in response and are unwilling to replace coffee trees with other crops. Consequently, when the market is weak, it can take a long time for coffee supply to adjust and prices to recover.

This instability is made worse by the fact that world prices are determined each day on futures and options exchanges in London and New York City. One function of futures markets is to attract speculators, who offset the physical purchases and sales of producers and users of the commodity. Speculators make money based on price fluctuations, exaggerating price volatility in both directions.

At these low prices, it is almost impossible for farmers to profit from coffee. Two recent related developments have contributed to this problem: the first is massive oversupply and the second is a widening gap in the market power of coffee farmers, at one end, and international trading and roasting companies, at the other. Coffee production is at its highest level ever. Between 1997–1998 and 2001–2002, coffee exports increased by 13 per cent (from 78 million to 89 million bags); but their total value fell by more than half, from US$12 billion to $5.6 billion (International Trade Centre, 2002, Table 1, p4).[2] The average coffee farmer produces and sells less than 15 bags (each 60kg) a year. By contrast, the largest roasters each purchase approximately 15 million bags of coffee annually and are able to capitalize on economies of scale and flexibilities afforded by their membership in huge companies such as Nestlé (the largest food company in the world) and Altria (known until early 2003 as Philip Morris).

The purchasing practices of large international roasters and traders have benefited from the increased export capacities of the major producing countries. Nestlé, for example, made an estimated 26 per cent profit margin on instant coffee, whereas Sara Lee's profits on coffee are running at an annual rate of nearly 17 per cent (Oxfam, 2002). Roasters are now able to concentrate their sourcing on a small handful of large producing countries rather than with several dozen smaller ones, and buyers have sent out powerful signals that they can absorb larger quantities of lower-quality coffee. Some of the largest roasters are establishing their own coffee plantations to extend their reach, reinforcing their control over the supply chain. If such trends continue, the market's diversity, with varieties grown as far afield as Java, Ethiopia, Jamaica and Guatemala, could be endangered. The situation threatens to erode farmers' livelihoods, as well as the economies of small countries that rely on coffee for half to three-quarters of their export income.

Compounding the effects of this instability, the market is now engaged in a restructuring process that is likely to favour large producers and big companies at small farmers' expense well into the future. The roasting and trading sectors have become more concentrated, with the leading four companies in each sector accounting for more than 40 per cent of coffee sales worldwide.[3] Their growing market power is reflected in their rising share of coffee incomes. While there are many elements that determine the cost of a jar of coffee or a cup of espresso, the basis of the consumer price remains the coffee bean. However, the share of the price received by the bean grower has steadily declined in the face of growing retail power and there is a widening gap between producer incomes and final prices. Since the fall of the International Coffee Agreement, for example, producers' share of the final retail price has fallen from 20 to 13 per cent (Talbot, 1997). This inequity is illustrated by the case of one small farmer, Peter Kafuluzi, who has grown Robusta coffee in Kituntu, Uganda, since 1945 and in 2003 sold the 'cherries' that contain green beans for just US$0.06 or $0.07 per kilo. The instant coffee into which they are made sells in the UK on a weight-for-weight basis for US$10 per kilo; yet the 15 people in Peter Kafuluzi's family receive less than 1 penny for every pound of coffee bought in a London supermarket (Oxfam, 2002, pp22–24).

Over a long period, this has created a ratchet effect, with retail prices adjusting more when bean prices have risen than when they have fallen. For example, the UK market is dominated by instant brands, largely made from Robusta coffee. Robusta prices reached their highest point during the last 20 years in January 1986, and then fell by 86 per cent to their lowest point in October 2001. Yet, over the same period, UK retail coffee prices increased by nearly one-third.[4] Similarly, while average green coffee prices[5] fell by 77 per cent from May 1997 to September 2001, retail prices in the US decreased by only 24 per cent.[6] Meanwhile, farmers like Peter Kafuluzi, who sell coffee to the large roasters, cannot afford enough food for their families or cannot keep their children in school (Oxfam, 2002, p26, footnotes 43 and 44).

The losers

The losers in this restructuring can be found in the poorest countries, with African, Andean and Central American countries among the hardest hit. Between 1992 and 2002, Brazil and Viet Nam's joint share of world production rose from 32 to 47 per cent (ICO, www.ico.org/prices/po.htm). This 15 per cent increase directly corresponds to the decline in Africa's market share, decreasing from 30 to 15 per cent over the last 30 years. The depth of Africa's coffee crisis is often overlooked. Yet, the significance of coffee to African economies is much greater than that of many coffee-producing countries. For example, as a small and poor country, Burundi accounts for just 0.3 per cent of world coffee exports; but during the late 1990s, coffee generated 80 per cent of its export revenue. And in Uganda, the livelihoods of roughly one-quarter of the population are in some way dependent upon coffee sales (Oxfam, 2002). Similarly, in southern Ethiopia, a country hard hit by food shortages, coffee accounts for over 50 per cent of export revenues and was critical to farmers during the great famines of 1973 and 1984. Yet, the impoverishment experienced in Africa is replicated in all corners of the world. In Central America, many coffee farmers have been forced to emigrate to cities or northwards to Mexico and the US, and the conditions in some coffee-growing areas have become so bad that children are suffering severe malnutrition. Similarly, there have been civil disturbances in Honduras and Guatemala, as well as reports of farmers' suicides in India (Oxfam, 2002, pp23–24, 30). Such scenarios are increasingly prevalent in the coffee trade, where farmers from Latin America to Asia face exacting constraints to ongoing production and market participation.

No easy road for farmers

Coffee farmers like Peter Kafuluzi have few options. They can stay with coffee, hoping for an upturn in the market and risk losses for years to come, or they can uproot their trees and face the uncertainty of other crops and unfamiliar new markets. If they opt to stay in coffee production, they need to either reduce their costs of production or increase their earnings by entering limited niche markets. Both options pose significant risks.

Farmers can reduce their output by cutting down trees or by allowing production to decline, using fewer inputs such as fertilizer and employing less labour to pick the harvest. The latter choice can have negative consequences. In many places, the quality of beans suffers since farmers harvest less frequently and more of the crop is unripe, over-ripe or in other ways imperfect. Selling lower-quality coffee, in turn, further reduces the price that farmers can obtain.

Some farmers who seek more secure and remunerative outlets for their produce have found new opportunities in differentiated niche coffee markets (see Fitter and Kaplinsky, 2001). While there has always been a system of discounts and premiums against standard prices for coffees of different origins, special premiums can now be obtained for coffees that meet specific requirements, such as shade-grown, organically grown and single-estate produced, while the fair trade movement provides an assured income and other benefits to farmers associated with it. This so-called specialty coffee sector has grown rapidly, now accounting for about 40 per cent of the value of US coffee market sales (Oxfam, 2002). These coffees attract the best prices – roughly US$1 per pound above the normal mild Arabica price in the case of Kenya AA and Guatemala Antigua to US$16 per pound for Jamaica Blue Mountain. As more attention is paid to them, some premiums are being pushed even higher (Lindsey, 2003).

For many farmers, obtaining these premiums may be the only way out of the crisis. Evidence from other markets, such as wine, Scotch whisky and table water, suggests that the share of premium coffees in the total market will continue to grow steadily. However, coffee farmers can benefit from this growth only to a limited extent. Expanding the niche market is a slow process, and it could be 20 years before niche products, taken as a whole, reach 20 per cent of the total market. Robusta growers are likely to be excluded as the premium requirements generally apply to Arabica coffees alone. The larger problem lies, however, in a so-called 'fallacy of composition': since demand for premium products is limited by their very nature, the premiums will shrink if too many growers try to enter the same market. Farmers will thus be facing the same situation of oversupply and declining prices that have led them to seek alternative markets in the first place.

Furthermore, large roaster companies are entering the niche markets despite the fact that they have previously shunned many of them (such as fair trade). Nestlé's Nescafe Partners Blend, for example, is the first instant coffee from a multinational to be awarded the Fairtrade mark. This could benefit the niche market segment by introducing it to a wider audience or devalue the niche brands in the eyes of traditional supports. At best, niche markets offer a partial solution to the coffee crisis, and for some farmers, they offer no solution at all.

Many commentators argue that the extent of global oversupply, coupled with market changes, offer farmers little choice but to cease production. They recommend programmes to help farmers develop different products and enter new markets as the main international solution. Crude diversification is already happening. Some coffee farmers are leaving the land; others are chopping down trees to grow other crops; still others are intercropping between the coffee trees. Yet, agricultural diversification is notoriously difficult. Some alternative crops to coffee, such as grains, are sold locally and depend upon coffee farmers' own prosperity to provide a market. With no established market access to even local markets, farmers will have to establish these links to markets themselves. Export crops such as cocoa, tea and export vegetables depend upon access to Northern markets as well as the capacity of growers to meet rigorous quality and technical standards. Other crops such as groundnuts are so heavily subsidized in Northern markets that poor farmers cannot compete on price. Even when market access is secure and production is lucrative, farmers may be engaged in the production of illegal crops such as coca, which has replaced coffee in parts of South America.

Even though large multinationals such as Nestlé and Kraft are assisting farmers in their supply chains to diversify, there is an urgent need for a more coordinated and coherent approach to diversification worldwide. Farmers would benefit from assistance in selecting the alternatives to coffee best suited to local soils, climates and market possibilities, and extension advice on production and marketing arrangements to help them to obtain optimal prices. Ultimately, a successful diversification programme should be designed by farmers themselves, serving their own needs on their own plots of land.

Roasting the roasters

The nature of the coffee crisis is complex and its solutions are correspondingly difficult. In order to confront the complexity of the issue, Oxfam developed a Coffee Rescue Plan aimed at bringing global supply in line with demand and supporting rural development so that farmers can earn a decent living from coffee. However, Oxfam acknowledged that the Coffee Rescue Plan will work only if all participants in the coffee market are actively involved. To this end, Oxfam developed specific challenges for all the major stakeholders – roaster companies, coffee retailers, governments (producer and consumer), multilateral institutions, consumers and investors.

Although the trade and marketing of coffee is diverse, Oxfam focused on the major multinationals, such as Nestlé, Kraft, Procter & Gamble and Sara Lee, whose brands (Nescafe, Maxwell House and Folgers, to name a few) have significantly shaped consumers' perceptions of drinking coffee. Why did Oxfam focus on these multinationals, dubbed 'roasters'? First, the roasters were very powerful: they had created international brands that defined the image and taste of coffee for hundreds of millions of consumers throughout the world. Their impact has been profound. For instance, every second 3900 cups of Nescafe are drunk in more than 120 countries (Oxfam, 2002, p6). Roasters also have been highly competitive with one another in most markets. If one roaster were to promote a new idea (e.g. fair trade coffee), the others would typically follow suit. The market power exercised by roasters is particularly important in the coffee market, which has been largely deregulated and where governments and international bodies have only a limited capacity to act. Second, the roasters were among the very few players in the market with the resources to act. These companies are fiscally strong, generally surpassing other food and beverage industries in revenues and profit margins. Third, the crisis was not in the roasters' long-term interests. Aside from the impact on their brands, several roasters acknowledged that the crisis was having or would have a serious impact on quality due to the inability of farmers to properly care for their crops. Although new technologies allowed roasters to absorb more low-quality coffee without damaging overall taste, they still required large volumes of good coffee. The largest roaster, Nestlé, cited ensuring a stable and reliable supply of good-quality coffee for the long term as a key consideration and shared Oxfam's concern that the coffee crisis could affect the quality of coffee bought by Nestlé (Oxfam, 2002, p43). Finally, the roasters were not neutral bystanders in the changes occurring in the coffee market and have been partially responsible for the difficulties that small farmers face. New technologies for blending and processing, for example, reduced the dependency of roasters on particular sites of production. This has made it difficult for coffee growers to differentiate their products from other locations and diluted their competitive advantage.

An important part of Oxfam's campaign was to convince roasters to conduct business differently. Oxfam had to convince them that their consumers cared about where their coffee was grown and how it was produced despite the fact that unlike other products, coffee blends do not usually display information about their origins or type (e.g. mix of Arabica/Robusta or country of origin). Nor do roasters divulge such information, or the price at which they purchased the coffee, to their suppliers. Oxfam challenged the roasters to commit to the following points:

- Ensure that the company knows how much the primary producer was paid for the coffee he/she sold. Commit to paying decent prices that enable farmers to cover their basic needs for their children's education, medicines and food. Make this information available.
- Support the fair trade movement by buying at least 2 per cent of total volumes from fair trade sources.
- Commit significant resources to tackle the coffee crisis – including financial contributions to diversification and aid.
- Lobby the US government to rejoin the International Coffee Organization.
- Support producing countries' efforts to curb overproduction by signing up to the quality scheme devised by the ICO. This scheme asks producing countries not to export coffees below certain grades, thereby reducing the amount of coffee on the market and helping prices to rise. A commitment from the roasters to only buy coffee above certain grades would strengthen the proposal.

Oxfam's focus on the supply chain was anathema to the coffee roasters. After all, they count their suppliers in millions. Furthermore, in between these suppliers (smallholders farming an acre or less of coffee trees) and the four major roasters lie a myriad of intermediaries: local buyers, millers, exporters, traders and importers. This makes it very difficult for roasters to establish transparent codes of conduct or benchmarks that can be monitored. For instance, no one could enforce a minimum wage equivalent for 25 million coffee farmers throughout the world. This lies at the heart of the challenge. If such a minimum wage were implemented, companies could claim that smallholders are effectively self-employed business owners and, as such, independent agents in the trading process, rather than employees who deserve special protection.

Despite the supply chain's complexity and fragmentation, Oxfam argued that roasters have a responsibility to coffee farmers. After all, consumers are sensitive to the abysmal working conditions in textile or toy factories that supply big brands in the US, even though the US retail or toy companies rarely own these factories. The supply chain is complex in both cases. But to the consumer, the issue is fairly clear: the person at the heart of the process – making the clothes, growing the coffee, picking the cotton – was not making money. In the case of coffee, it was worse. The growers were actually losing money because they were growing coffee below costs of production, effectively subsidizing the profits of the multinationals.

Furthermore, the coffee could be traded differently even though fragmentation makes transparency difficult. Movements for fair trade and organic coffee have shown that certification, and rules of origin and labelling, are not only possible but commercially viable, given consumers' growing interest in the coffee production process. For the roasters, however, such a change would dramatically affect how they do business, and it would extend ethical sourcing to a new market. Unlike workers in manufacturing sectors, for instance, the millions of farmers exporting cash crops are largely unprotected by national or international conventions and regulations governing the well-being of the labour force.

Somewhat to Oxfam's surprise, some companies were ready to acknowledge their responsibility to coffee farmers. Nestlé, for instance, declared that:

> *A few years down the road, we are going to be asked not only if we have maxim-ized short-term shareholder value, but also some other, more difficult questions. Among them will certainly be: what have you done to help fight hunger in developing countries?* (Oxfam, 2002, p26)

Oxfam attempted to capitalize on the willingness of companies to address their social responsibility, requesting that they adopt the Fairtrade standard for ethical sourcing in coffee in order to guarantee farmers a higher than world market price for their coffee. In the UK, for instance, fair trade coffee accounts for over 7 per cent of sales of roast and ground coffee (Oxfam, 2002, p42). Nearly 200 coffee co-operatives, representing 675,000 farmers and about 350 coffee companies, adhere to the standards of the Fairtrade Labelling Organizations International (FLO) to bring products to market in a way that ensures that farmers receive a decent return. The hallmarks of fair trade coffee are as follows:

- Farmers are guaranteed a certain price. For Arabica, the vast majority of fair trade coffee, this price stands at US$1.26 per pound, well above the market price as traded according to the New York 'C' contract (New York is the centre of coffee trading – electronically). In addition, organic coffee received an additional 15 cents per pound premium.

- Fair trade principally works with small-scale coffee farmers associated through co-operatives, with an emphasis on democratic governance.
- Farmers receive benefits from fair trade in addition to a better than market price – for example, pre-financing, which can reach up to 60 per cent of the contract's value. The fair trade movement also stresses the need for technical and business training so that co-operatives can capture more of the value added to the products they export.

Fairtrade does not address all elements of sustainable agriculture. For instance, Rainforest Alliance, another longstanding ethical certification organization, is stronger on environmental and workers' rights issues and working with small-scale farmers not organized in co-operatives. But the higher price guarantee of Fairtrade dealt with one of the key issues of the coffee crisis. Other challenges for Fairtrade include a rigid business model that limits the opportunity to sell Fairtrade-certified coffee outside the dominant UK and European markets; financial costs that farmers must carry to enable them to be certified by Fairtrade; and certifying only the raw commodity and not the manufacturing process itself. Notwithstanding these limitations, Fairtrade does address a key element of the coffee crisis – the low price farmers receive.

Although Oxfam asked companies to pay farmers a decent price, it also pressured roasters to change the way in which they conducted their business. In addition to the concrete proposals outlined above, Oxfam emphasized the following:

- Farmers need more predictable incomes, either through the greater use of fixed contracts or letters of intent for future supply agreements.
- Companies, both roasters and traders/importers, need to work with co-operatives and associations of farmers on technical issues. In many of the poorest countries, crucial extension services have been dismantled, and unless companies fill this vacuum farmers will remain unaware of market demands and how best to address them. Technical skills, ranging from pest control to milling/hulling and good husbandry (when to pick the coffee cherries, for instance, and how quickly to process them) would improve farmers' negotiating position and motivate them to produce for value, not volume.
- Companies need to reward farmers for crop quality. Although the green bean that is traded internationally is priced according to quality criteria relative to the benchmark coffee, the original producers rarely see that premium (or discount) because intermediaries rarely take the quality of coffee into account. This may be because the exporter does not reward them for it or because the coffee that the farmers sell is not processed sufficiently and thus its quality cannot be ascertained.

While ethical sourcing can contribute to alleviating some aspects of the coffee crisis, it is not enough. Improvements in the companies' supply chains must be accompanied by structural changes in the market that address the issue of oversupply. Oxfam highlighted several plans, such as the International Coffee Organization's quality scheme, which Oxfam urged roasters to support, and developed additional proposals to address oversupply and other systemic issues. These included the following:

- An international forum was created, bringing together coffee companies, producers, non-governmental organizations (NGOs) and other stakeholders to establish a common code for coffee (see www.sustainable-coffee.net). This common code is still being elaborated. While it makes no claim to improve coffee prices directly, the code aims to improve farmers' livelihoods through more efficient production.
- Starbucks' 'preferred supplier' scheme, or Café Practices, which pays farmers a premium for complying with general environmental and social standards, was updated – in certain cases, Starbucks pays an even higher price than Fairtrade. These Café Practices are the leading standard in company certification and in many ways are an improvement on Fairtrade and other ethical certifiers. For instance, it is the only certification that includes the criteria of price sustainability transparency to ensure that farmers receive a sustainable price.
- Nestlé piloted a Fairtrade-certified Nescafe coffee called Partnership Blend in October 2005. Not only is it the first major roaster to sell a Fairtrade-certified coffee, but it also focused on the origin of coffee by sourcing and promoting Ethiopia and El Salvador as the coffee's origin. This development has the potential to revolutionize the coffee market in a positive way in the same way as Starbucks' Café Practices have done – by expanding the certification and origin of coffee into the mainstream coffee market, an area that Fairtrade has traditionally struggled with.
- Procter & Gamble committed to purchasing fair trade coffee for its Millstone brand and to buying certified shade-grown coffee through the Rainforest Alliance.
- The Co-operative Group became the first UK supermarket to switch all of its own-label coffee to fair trade. It is expected to add about UK£4 million a year to fair trade coffee sales in the UK – an increase of 15 per cent. It also will triple the Co-op's existing sales of fair trade coffee.
- Kraft did not commit to fair trade, but did sign an agreement with the Rainforest Alliance. This agreement committed Kraft to buying 5 million pounds of coffee in the first year. This coffee will be certified as sustainably managed by the Rainforest Alliance. Although the focus is not on price, farmers do benefit from increased productivity, efficiencies and cost reductions due to improved farming techniques. Furthermore, the Rainforest Alliance is especially strong on environmental and workers' rights issues, and has a well-defined methodology in helping farmers to improve production processes and quality.

- Dunkin Donuts committed to buying fair trade coffee for a new espresso beverage and expects fair trade coffee to represent about 2 per cent of the company's total coffee purchases. This will result in approximately 1.2 million pounds of fair trade coffee being purchased.
- Starbucks increased its fair trade coffee purchases to 4.8 million pounds in 2004 – up from 2.1 million pounds in 2003. At present, it is the single largest buyer of fair trade coffee. It also pays the highest price for its coffee outside fair trade – an average of US$1.20 to $1.31 for its coffee, which also benefits farmers.
- Other selected companies and organizations that have switched to fair trade or have started supporting fair trade include Astra Zenica; the British Film Industry; the UK Department for International Development (DFID) and the Department of Trade and Industry (DTI); the Eden Project; the European Commission (selected canteens); Evershed; the Greater London Authority; the House of Commons in the UK; Merril Lynch; Microsoft Nationwide in the UK; Orange; the Scottish Parliament; UK universities (Oxford Brookes, Birmingham, Edinburgh, Nottingham, Bristol); over 50 cities and towns in the UK; over 200 college campuses in the US (including Harvard, Yale and Brown); 5000 US faith-based congregations; the World Bank; the US Senate and Congress; and McDonalds in Switzerland.
- Sales of fair trade coffee increased globally from 10 per cent in 2002 to 19 per cent in 2003.
- Individual companies have also contributed in other ways. For instance, Nestlé increased its direct purchasing from 12 to 14 per cent in 2003; it supported the ICO Quality Scheme and lobbied the US government to rejoin the ICO. It has also actively worked with governments and other industry officials to develop common action plans to address problems.

Despite these improvements, the coffee crisis persists, and Oxfam continues to pressure all stakeholders (especially those in the private sector) to face the problems with greater commitment. Oxfam has subsequently challenged stakeholders to identify a workable solution that will require industry, governments and producers to work together to address the coffee crisis. To this end, Oxfam continues to pressure companies to improve their supply chains and participate in a workable strategy to limit coffee oversupply.

Three years after Oxfam's campaign launch, coffee companies such as Nestlé, Kraft and Starbucks are leading the way in increasing the pay that farmers receive for their coffee. Fairtrade and the Rainforest Alliance are also working with these companies in providing new and more sustainable ways of operating in the coffee industry. The challenge, now, is to broaden the work of these companies and certifiers to ensure that scale and equity is reached throughout the coffee industry. This, perhaps, is the largest challenge of all.

Conclusions

Small coffee producers continue to face precarious conditions, struggling to maintain production in the face of low world commodity prices, corporate consolidation and trade liberalization. For many of the world's 25 million coffee farmers, the coffee crisis has become a humanitarian crisis that requires urgent action. Oxfam has attempted to address this situation through a multi-pronged Coffee Campaign that aims to restore the balance of supply and demand, increase prices and revive livelihoods, and establish viable alternatives for rural development. Yet, the success of this strategy rests on engaging the commitment and active participation of all stakeholders in the industry, from governments to global corporations, in the process of making trade fair. Companies such as Nestlé, Kraft and Starbucks, and certifiers such as Fairtrade and the Rainforest Alliance, are leading the way in changing the coffee industry's approach; but a reverse in the current crisis will only occur if the coffee industry can serve the needs of the poor as well as the rich – and this challenge needs more than the individual actions of companies and certifiers.

Notes

1   This is the form that is normally traded between countries. It consists of the beans that are extracted from coffee cherries and that are ready for the roasting stage in processing.
2   Over the same period, production increased from 96 million to 110 million bags; but consumption increased from only 103 million to 107 million bags, turning an annual deficit of 7 million bags into a surplus of 3 million bags. As a result, by 2002 there were stocks of more than 40 million bags overhanging the market (data from the International Coffee Organization, www.ico.org/frameset/traset.htm; the United Nations Food and Agriculture Organization, www.fao.org/es/ESC/esce/cmr/cmrnotes/CMRcofe.htm; and J. Ganes Consulting, www.jganesconsulting.com/samples/In-Depth_Coffee_Report.pdf).
3   This question is discussed in more detail in Oxfam (2002, pp13–14).
4   As expressed in US dollars, from US$9.17 per pound to US$11.92 per pound (ICO, www.ico.org/prices/po.htm).
5   As measured by the ICO's composite indicator.
6   As measured by the ICO's composite indicator. The calculations of the decline in prices are the author's.

References

Fitter, R. and Kaplinsky, R. (2001) 'Who gains from product rents as the coffee market becomes more differentiated? A value chain analysis', *IDS Bulletin Special Issue on the Value of Value Chains*, vol 32, no 3, pp69–82
International Trade Centre (2002) 'Coffee: An exporter's guide', International Trade Centre, UNCTAD/WTO, Geneva, Table 1, p4

Lindsey, B. (2003) 'Grounds for complaint? Understanding the "coffee crisis"', *Trade Briefing Paper*, no 16, 6 May, www.freetrade.org/pubs/briefs/tbp-016.pdf

Oxfam (2002) *Mugged: Poverty in Your Coffee Cup*, Oxfam, www.maketradefair.com

Ponte, S. (2002) 'Standards, trade and equity: Lessons from the speciality coffee industry', *Centre for Development Research Working Paper*, vol 2, no 13, November, Copenhagen

Talbot, J. M. (1997) 'Where does your coffee dollar go? The division of income and surplus along the coffee commodity chain', *Studies in Comparative International Development*, vol 32, pp56–91

10

Small Producers: Constraints and Challenges in the Global Food System

Tom Fox and Bill Vorley

Introduction

Small-farm agriculture has been presented as a growth-equity 'win–win', and this has encouraged a resurgence of interest in smallholder agriculture in the poverty reduction debate. But making small farms effective agents of wealth creation is quite a challenge for governments during a period of globalization and market liberalization. The opportunities for smallholder farming may be restricted by the costs of dealing with modern markets. There are concerns that the evolving structures of food chains, characterized by the concentration of market power in the hands of large processors and retailers, and the new standards associated with modern chains, can work against the interests of small- to medium-scale producers and workers. This can be either through creating barriers to market entry or worsening the terms on which they engage in trade.

Private-sector strategies in the agri-food sector, especially in global retailing, are moving fast under the radar of public policy. We are heading into an era of growth of international supply chains, a reduced role for state organizations and recasting of regulatory systems, with a lack of information and mechanisms for effective policy response. If policy is to anticipate rather than lag behind these changes, then those changes – and their implications for rural producers – must be better understood.

This chapter examines the issues associated with the entry of small-scale agricultural producers into modern agri-food chains, with the features of changing power relations in food chains between producers and 'downstream' players, particularly the supermarket sector. We discuss the *private re-regulation* of agriculture and the ability of buyers to set product and process standards as entry tickets to their supply chains, comparing bulk commodity markets and 'buyer-driven' chains. We then describe the areas of policy influence – both public and private sector. We examine the limitations of 'niche' options, such as organic and fair trade. In contrast, we put the case for 're-governing' market structures in favour of new business opportunities for poor rural producers and workers as a means of achieving more equitable rural development and fairer trade.

Bulk commodity chains

Bulk commodity chains deal in undifferentiated staples, such as the majority of trade in wheat, soya, coffee, palm oil, cocoa and sugar. These commodities are usually bulky and storable. There are few buyers and many sellers, and trade is characterized by flexible sourcing from diverse locations. Marketing is at arms' length at central spot markets, and price determines when and where the product moves. Trade is based on anonymity and standardization, which keep information flows between trading partners to a minimum. A small number of often privately owned companies control key elements of production, trade, processing and marketing.

The undifferentiated nature of these markets means that it is easier for small and family-scale farms to participate. But bulk commodity markets are characterized by instability, structural oversupply, stiff global competition, historic downward price trends and declining terms of trade for producing countries and regions. In producer countries, the privatization and liberalization of commodity exports such as cocoa make it more difficult for countries to control the flow of exports and, thus, influence world prices.

Increasingly, it is companies based in industrialized countries that are capturing value-added onto tropical bulk commodity products through branding and re-exportation. There is a widening gap between world prices for agricultural goods and retail prices, and this gap has accelerated since the 1980s. The developing country contribution to value-added in the cocoa sector, for example (measured as value of exports of cocoa beans, cocoa products and chocolate), declined to around 28 per cent during 1998–2000, down from around 60 per cent in 1970–1972 (Östensson, 2002). Retail prices for coffee have remained relatively stable despite producer prices dropping to less than one-third of their 1960 level. This has fuelled accusations of profiteering from the impoverishment of millions of smallholders. According to a recent United Nations Conference on Trade and Development (UNCTAD) roundtable, annual export earnings of coffee-producing countries during the early 1990s were US$10–$12 billion and global retail sales were about US$30 billion. Now, retail sales exceed US$70 billion, but coffee-producing countries receive only US$5.5 billion. A World Bank report (Morisset, 1997) estimated that divergence between producer and consumer prices may have cost commodity-exporting countries more than US$100 billion a year.

At the producer end, the withdrawal of the state from direct involvement in commodity markets and the abandonment of international commodity agreements expose producers and labourers to price fluctuations, without the traditional safety nets of credit and state trading institutions. Relocation of risks from the state to the individual means that farmers now bear the opportunities and risks of direct exposure to volatile and unpredictable commodity markets.

Buyer-driven chains

To escape from the low prices and intense competition of bulk commodity markets, there is much interest in opportunities for small farmers to 'upgrade' to coordinated 'buyer-driven' chains (Gereffi, 1994). This is because buyers in these chains are willing to pay a higher price for differentiated, de-commodified products and may provide better support services. Much attention has been focused on the organizational, technical and institutional arrangements by which small producers can deal with the requirements of buyer-driven chains.

Buyer-driven chains are found where product uniformity and high quality are necessary for further processing, branding and retailing. They arise from a need for specific processing traits (such as feed quality or starch quality of grains), as well as assurance of supply, traceability and preservation of identity. Traceability demonstrates 'due diligence' and manages risk, especially against contamination of food by pathogens (e.g. BSE and *Escherichia coli*), toxins (e.g. pesticides and dioxins) and alien genes (e.g. Starlink). Possibilities of widening the set of product attributes to include sustainability of production, processing and handling have opened up; the application of management systems for environmental (such as ISO 14000) or production system (e.g. organic) can be preserved along chains of custody. The high end of commodity markets – from gourmet coffee to identity-preserved grains – is also now moving in this 'de-commodified' direction.

It goes without saying that buyer-driven chains are more regulated than bulk commodity markets and are characterized by higher levels of private-sector governance and long-term coordination between producers, suppliers, processors and retailers. The degree of buyer governance varies considerably; but in the most extreme examples, such as pre-packed fresh vegetables, production sites around the world may be visited frequently by retailers to ensure compliance with their codes and standards. Informal standards have emerged as a key tool to manage quality, food safety and various intangible attributes relating to production practices within the supply chains of supermarkets and branded manufacturers and processors.

Buyer-driven chains bring about market segmentation, which means that producers are contracting more actively with their customers – the processors and retailers – in order to deliver differentiated products. Contracts cover such parameters as quality, quantity and price premium. Production contracts can improve coordination and efficiency, allowing a company to influence production, reduce procurement costs and price risks, and maintain flexibility while avoiding the risks and capital associated with farming. Contract farming can bring significant benefits to producers; a farmer is assured of a buyer, price risk is reduced, favourable credit terms may be available, and marketing costs are lower. The 'reversal of the marketing chain' from a production to buyer orientation can also benefit consumers. It is no coincidence that in the UK, where supermarket power is most ascendant, consumers' aversion to genetically modified (GM) technology was translated into retailer-driven programmes to purge own-brand supply chains of GM ingredients.

Supermarkets and buyer-driven chains

The major force in the development of buyer-driven chains and the 'private re-regulation' of agriculture is now viewed as the retailers, especially global supermarket companies. The modern supermarket model is an *accelerator of structural change* towards vertical coordination of agri-food chains. It is linked to a key feature of the modern supermarket model: marketing strategies built around trust and the defence of quality, consistency and assurance to consumers through traceability systems, especially in support of supermarkets' private brands. Supermarkets endow themselves and their brands with private standards and certification schemes, which fill an institutional gap left by public regulation. Thus, supermarkets can pick up custom with every food scare and animal health crisis, as seen in China with the severe acute respiratory syndrome (SARS) outbreak. Retailers' own brands return the highest contribution margin or gross profit and 'have been one of the competitive forces which shifted strongly in favour of retailers' (Wrigley and Lowe, 2002).

The position of farmers in supermarket-driven chains was considered of interest only to industrialized world farmers and exporters to the North. But supermarket dominance of agri-food is no longer simply an industrialized world phenomenon. The growth of supermarkets in the South means that primary producers and processors face domestic markets that start to take on the characteristics of export markets.

Ground-breaking work in Latin America and elsewhere has shown that penetration of transnational retail firms is proceeding at a rapid pace even in rural areas of the developing world, and this is having a marked impact on market structure and governance (Reardon and Berdegué, 2002). Just about all population growth over the next 25 years is predicted to take place in urban centres in low- to middle-income countries, and global retailers are structuring their organizations to follow this location of demand. Supermarket companies have long since moved out of the strategy of serving only the middle classes and expatriate populations. Carrefour, for instance, the most successful hypermarket operator in China, has a stable consumer base in China among people of low- to medium-income levels and uses different formats, such as the Dia discount format, to develop the lower-income market segments. Supermarket chains, once established in the richer/larger regions and countries, seek competitive territory in the smaller or poorer countries. Local and regional players, such as the Bailian group in China and Shoprite in Southern Africa, may also have developed skills in working in risky environments and low-income consumer segments – for instance, through franchised convenience retail and wholesale formats.

The inevitability of retail's dominance by 'modern' supermarket formats is not a foregone conclusion. It is clear that the traditional agri-food actors do not stand still, but learn from and respond to these changes, leading to forms of co-existence. But this adaptation to change is itself a driver of restructuring at other levels. Global cash-and-carry wholesalers such as Metro can supply small retailers, food service companies and institutional buyers at prices below those of the traditional outlets because of their buyer power and knowledge in packaging, labelling, product specification, logistics and infrastructure. While the front end of independent retailing in countries such as Viet Nam may look unchanged, supply chains will operate very differently when wholesale is dominated by global companies, resembling the integrated chains of the global retailers and providing wider channels for food imports and exports. Metro is one of the most international grocery retailers, active in 28 countries, with seven markets entered since 1999, predominantly in Asia and Eastern Europe. In India, where foreign direct investment in retail is prohibited, Metro Cash and Carry was permitted to set up two distribution centres in Bangalore in 2003 and has plans for a network of around 20 outlets, generating annual revenues of US\$1 billion per year in India within 5 years.

Market power and the distribution of costs and benefits

The most problematic element of the relationship between small producers and buyers may not be 'upgrading' to the demands of these new chains. In the medium to long term, the biggest challenge is the acceleration of the modern supermarket model – comprising discount pricing, large volume procurement and a centralized distribution system – towards a highly concentrated structure in which most power and leverage resides at the retail end of supply chains, and in which benefits are passed to customers and shareholders. From a macro perspective, this means less residual value shared with other actors in the chain and a shift in value from producer to consumer, and from countryside to city.

The supermarket firms are engaged in an increasingly fierce price war in which the winners are those who can (among other competitive factors) work their supply chains the hardest in terms of extracting savings and services from suppliers. Intermediaries such as the dairy processors or fresh produce integrators have grown and consolidated alongside supermarkets, and have managed to claw back some market power. Primary production, however, is the part of the chain where the exercise of market power and accumulation of value is most curtailed. Nowhere is this better illustrated than bananas, where the supermarket price wars have been costly for suppliers (see Box 10.1). The debate about small farmers and modern food chains is inseparable from the debate on corporate concentration.

Globally, concern is emerging that the concentration of economic power by industries along the chains between primary producers and consumers – the traders, processors and retailers – is affecting the profitability and livelihoods of primary producers and workers. This was underscored by a milestone statement on industrial concentration in the agri-food sector issued by the International Federation of Agricultural Producers (IFAP) in May 2002, which states:

> *Much attention has rightly been drawn to the distortions caused by certain types of government policies. However, relatively little attention has been paid to the market distortions caused by the high level of concentration in the input and distribution side of the agri-food system. Yet, it is clear that the domination of a few large firms, both upstream and downstream of the farming sector, can significantly affect market conditions.*

Market share is the traditional measure of supermarket success in the marketplace. A higher market share allows economies of scale and the extraction of better terms from suppliers. Evidence that large buyers can extract more favourable terms from suppliers – through bulk-buying economies, through playing suppliers off against each other or through threats of de-listing – is not hard to find. The UK Competition Commission's (2000) report on supermarkets clearly shows that the largest supermarket, in this case Tesco, can consistently obtain discounts from their suppliers 4 per cent below the industry average, while the smaller players pay above the odds. With retail margins often quite low, these differences in supplier prices have a profound impact on supermarket profitability and are a frank demonstration of the link between size and buyer power.

Increased market share and sales density deliver lower unit costs and higher net margins, potentially leading to a 'spiral of supermarket growth' (Burt and Sparks, 2003). Features of this spiral are that absolute costs and barriers to entry for competitors are raised, and growth becomes dominated by one or two organizations. The savings from the growth spiral can be invested in facilities, or in more sustainable supply chain operations, or can be passed on to consumers in lower prices. The success of Wal-Mart shows that investing the revenues of the growth spiral in lower consumer prices in order to capture larger market share has been a successful strategy.

One expression of market power is the allocation of the costs and benefits of production standards. These include the EurepGAP protocols, designed by a group of European food retailers primarily with food safety in mind, but with some reference to social and environmental issues, and the many individual sets of standards which companies have developed. Although often labelled as 'voluntary' standards in that they are not imposed by regulatory authorities, such requirements often act as 'entry tickets' into the market – producers must

Box 10.1 The global banana chain: The impact of buyer power

The global banana chain provides an example of the impact of buyer power on primary producers within a market, which, since the 1990s, has been characterized by oversupply, weak prices and increased competition between distribution companies.

The global trade in bananas is a classic oligopoly. While a portion of trade is in the hands of the companies of independent national growers, importers and ripeners, a small number of vertically integrated transnational corporations dominate international banana marketing and trade, and, according to the United Nations Conference on Trade and Development (UNCTAD), 'are able to exercise their market power at several or all the stages of the banana marketing chain'. Only around 12 per cent of revenues from banana retail sales remain in producing countries despite the very limited amount of product transformation outside of the farm or plantation.

The dominance of retailers has had an increasing influence over the structure and distribution of value along the banana chain. The shift of profits downstream has been dramatic over the last decade, and the transnationals' margins on bananas are now very slim. Although these multinationals are vertically integrated in sourcing, shipping, ripening, packing and distribution, they are moving away from direct ownership of production. As with other commodities, preferred supplier arrangements are now the norm, with contracts specifying standards for quality, packaging, etc. (Fajarnes-Garces and Matringe, 2002).

Bananas are the UK's most popular fruit and are a 'known value item' – that is, price awareness among consumers is high. When one leading supermarket drops the price of bananas, the rest are obliged to follow. A price war initiated by Asda–Wal-Mart in mid 2002 led to a reduction in supermarkets' supply bases and the culling of less competitive suppliers as all major supermarkets demanded deep price cuts at the supplier side. This meant that it was impossible for a grower in Costa Rica to be paid a legally minimum price for a box of bananas. In turn, it was impossible for that grower to pay labour a legal minimum wage. In Costa Rica, plantation workers' daily wages have fallen from around US$12–$15 in 2000 to around US$7–$8 in 2003 (Banana Link, cited in Vorley, 2003). Small-scale producers in the Caribbean are watching further consolidation in the UK supermarket sector with trepidation.

There are two pertinent lessons from the banana example for the debate about upgrading. First, the downward pressure on prices in the mainstream sector can undermine the 'quality' (organic, fair trade, etc.) niches by lowering the reference price from which consumers make purchasing decisions. Second, there is the risk that 'alternative' niches, such as fair trade – rich in 'success stories' and examples of 'corporate responsibility' – can act as a fig leaf over widespread injustices in trading relationships within mainstream supply chains.

Source: Banana Link, cited in Vorley (2003)

comply with the standards, and be able to demonstrate that they have done so, or their products will not reach the supermarket shelves. Technical assistance and extension services that support small farmers are very ill equipped to assist farmers in meeting the standards imposed by supermarkets.

Standards can bring producers considerable benefits, such as reduced agrochemical use and higher self-esteem. But they may also be regressive instruments with relative higher costs and complexity falling on the smallest operation. At issue is the share of costs and benefits between the standard makers and standard 'takers'. Large producers can spread the cost of inspections, audits or certification across their entire operations. By applying a 'one-size-fits-all' model, the standards may include inappropriate expectations for small and poorly resourced companies, and drive the rationalization of supply chains, with a few large suppliers preferred to many smaller suppliers. Furthermore, as markets mature, meeting standards is no guarantee of a market premium.

In the medium to long term, buyers, in effect, create *captive suppliers* without having to increase investment in the sectors. These 'dedicated' or 'preferred' suppliers and producer organizations, set up with the aims of enhancing traceability, quality assurance and developing closer links from the farmer through to the consumer, are faced with both the 'chain insider' benefits (such as being supported through hard times by a customer) and 'one buyer' risks of exclusive partnerships.

The UK milk sector is particularly illustrative. During 2004, all three top supermarket chains in the UK continued a process of drastic rationalization of their liquid milk supplies. Both Tesco and Sainsbury announced that they were rationalizing their supply base to only two milk suppliers. Asda–Wal-Mart made the decision to buy its milk solely from Arla Foods. Being a single supplier does not in itself mean that retailers are exerting specific price pressure, but it certainly 'creates conditions in which competition creates its own pressure' to the extent that retailers can cap farmgate prices against their targets for profit margins.

New research in Brazil (Mainville and Reardon, 2004) and Central America (see, for example, Hernández et al, 2004) shows that suppliers to supermarket chains have higher gross incomes; but these can be wiped out by the costs (such as increased use of inputs) of meeting higher-quality standards. Furthermore, the higher returns come under strong pressure as markets mature and there seems to be no guarantee of higher profits from retail-driven chains in the long term. But conforming to high standards for one retailer opens up new markets for growers; other supermarkets without their own standards will look favourably on producers that can deal with the industry leaders.

Policy implications: Options for action

Market restructuring is moving quickly. The risks for small producers in dynamic markets are very high, and opportunities for sustained participation in these markets are limited, undermining the comparative advantage of small farmers where it exists. Domestic markets even in the South are not necessarily a refuge as restructuring spreads in the wake of global and regional processor and retailer expansion into middle-income countries such as China, South-East Asia and most of Latin America. Nor are 'organic' or other 'sustainable' niches insulated from restructuring despite their popularity with donors. As Spoor (1997) notes, the 'high-profit, low-volume' markets that have emerged after liberalization are far smaller than promised by the designers of structural adjustment. It is not possible to 'niche market' the way out of a crisis in small-scale agriculture.

There is a lot of policy advice to farmers about 'adding value' and creating relationships with customers, although the distribution of 'value-added' is skewed to downstream processors and retailers. Some smallholders will, indeed, succeed in upgrading from the vagaries of bulk commodity production to higher value chains, with the right combination of organization, entrepreneurship and technical support. But for the majority of marginalized farmers in countries such as Bolivia, which are themselves very marginalized (and subject to huge inflows of cheap products from neighbouring countries), dealing with buyer-drivers is a very challenging prospect. The capacity for public policy responses is currently severely limited due to a lack of information and mechanisms for anticipatory policy development and planning in what is a dynamic, complex and moving agenda. Nevertheless, some areas of policy intervention are immediately apparent.

Cooperating to compete

The most obvious advice for small-scale and family farmers responding to the changes in agri-food organization is to treat the changes as the new commercial reality and to organize to engage with this reality. Small producers in both developing and industrialized countries are being advised to forge direct relations with the market, as well as with providers of research and advice, with non-governmental organizations (NGOs) and with the state. This is the logic of 'small farmer economic organizations' (SFEOs) in the developing world (Berdegué, 2001) and 'new generation' co-operatives in the industrialized world. These are producers who realize that in a chronically oversupplied and liberalized market, a marketing mentality – in which organizations perform at higher levels of specification – is necessary to contract into modern coordinated chains. A SFEO may be set up by producers around a common interest in generating improved income through the joint production and/or marketing of a commodity; accessing market information; minimizing marketing costs; unifying their production goals; improving quality; and possibly cutting out the middlemen and farmgate buyers.

SFEOs are a hot development topic because of their perceived link between small producers, the state, donors, and national and international markets. They are seen as being in a crucial position to combat poverty within a liberalized economy and as key to formalizing the smallholder economy. Participation in economic organizations can bring significant economic benefits when the organization operates in buyer-driven chains with high transaction costs, such as dairy (Berdegué, 2001). They may be well placed to deal with the management requirements of regulations and inspections associated with buyer-driven chains. Success depends upon group solidarity, collective bargaining techniques and institutions that enforce contracts impartially and that secure long-term property rights. But many SFEOs lack the capacity to define, monitor and enforce internal rules to be able to meet the conditions of modern agri-food chains.

A case study from Bolivia (see Box 10.2) shows that when governance of the chain changed towards a more 'buyer-driven' organization, the SFEO was unable to exert any significant leverage over the development of the milk market. There is a growing appreciation that the donor community has probably loaded an overly ambitious set of expectations on SFEOs as direct market actors in an era of deregulation and trade liberalization (Muñoz et al, 2004).

Box 10.2 'La Gloria' in Bolivia

The evolution of the milk market in Bolivia dramatically illustrates the impact of economic globalization on small farmer organizations since the privatization and subsequent capture by transnational capital of the state milk enterprise. Three public milk companies (PILs) were set up by the Bolivian state to supply milk to the major Bolivian cities during the 1960s and 1980s as social and economic enterprises. The PILs received millions of dollars of investments from the state and from international aid agencies. The three plants account for the vast majority of the country's industrialized dairy production.

The Association of Milk Producers in the Province of Aroma (ASPROLPA) was established in 1992 to coordinate the supply of milk from Aroma province on the high *altiplano* to the PIL near La Paz, providing social control of quality and supply. It was also to represent members organized in 'modules' in negotiations with the government on issues of price, credit and technical assistance for livestock development. At its peak, ASPROLPA produced 10,000 litres per day, equivalent to 30 per cent of milk production in the province, from areas of severe natural resource constraints and deep rural poverty.

The Peruvian food and construction conglomerate La Gloria bought a controlling stake in the PIL's Cochabamba and La Paz operations for US$8 million when they were privatized in 1996, followed by the purchase of the Santa Cruz company in 1999 for US$10.5 million. Conditions were attached to the 1996 privatization sale, in which La Gloria paid Bolivian milk producers a premium over prices paid to lower-cost producers in Argentina. The contract also required La Gloria to buy all of the milk produced by the modules until the end of 2001.

The market then began to take on classic buyer-driven characteristics. In September 2000, the PIL closed the La Paz processing plant and converted it to a distribution centre for milk arriving from the more efficient plant in Cochabamba. Milk produced on the *altiplano* now travels 600km to La Gloria's processing plant in Tacna, Peru. La Gloria asked all modules to install cooling tanks (at a cost of US$6000 each) so that collected milk would meet its new quality-related standard of 4°C. Only two modules had these tanks already installed, and the other communities clearly could not afford the investment. As an alternative means of reducing collection temperature, La Gloria then asked for collection during the middle of the night, at 2am, when milk temperature was low. When farmers complained, La Gloria threatened to abandon milk collection and to pass responsibility of milk delivery to the farmers.

The situation improved somewhat during late 2000 when the local ice-cream manufacturer Delicia entered the market. But, as Bebbington (2001) notes, the capture of the state enterprise's milk collection and processing infrastructure by private capital, and the lack of ownership in the newly privatized industry by ASPROLPA members, leaves the organization with very little leverage over the development of the milk market. Membership of the Southern Common Market MERCOSUR (*Mercado Común del Sur*, currently Argentina, Paraguay, Uruguay and Brazil) may open a floodgate of cheap milk from Uruguay and Argentina, and ASPROLPA is looking to its local market on the *altiplano* as a potential survival strategy.

At the Santa Cruz plant, milk prices have fallen for both the formal and informal producer sector since the PIL was sold to La Gloria. This price reduction was not passed to the consumer.

Source: Bebbington (2001); Muñoz et al (2004)

Re-regulation of commodity markets

Donor agencies, in their search for 'sustainable markets', are looking for the elusive 'win–win–win' of environmental protection, poverty alleviation and economic growth. The temptation is, then, to home in on micro-niches such as smallholder exports of organic fair trade produce from the developing world. But to focus on these niches is to duck the issue of reform of mainstream markets, be they bulk commodity markets, traditional markets or buyer-driven chains. In this regard, the first priority is to take stock of opportunities to *re-regulate commodity markets*, using progressive interventions that reduce volatility and avoid surpluses, as well as building in social and environmental objectives. This was the subject of intensive discussion during the 11th United Nations Conference on Trade and Development (UNCTAD) in 2004.

Investments in traditional markets

Traditional wholesale and retail markets have been largely ignored by policy-makers, and the lure of new supermarket chains risks a continuation of this neglect. Investment in traditional markets in areas of hygiene, standards, safety and infrastructure can give small-scale suppliers more options, and also helps those markets to compete with supermarket chains and make the transition to supplying supermarkets.

Better information on market structure

The debate about market restructuring and smallholder access to markets suffers from a poor information base, including the extent of corporate concentration in global food chains. How much of a transition from bulk commodity/staple production to buyer-driven chains is actually under way? How easily saturated are the new quality markets? Are some markets actually headed the other way – being commodified – as seen in the increased use of electronic auctions by supermarkets? Is the growing market share of discount supermarkets hastening the commodification process? What are the real impacts of supermarkets in the South relative to other drivers of agri-food restructuring?

Considering how much of agri-food trade, processing and retailing is in the hands of a small number of corporations, the case for monitoring transnationals at the United Nations level should be pursued as a matter of urgency. The role of the extinct United Nations Centre for Transnational Corporations (UNCTC) included information collection, research, policy advice and the development of standards of behaviour. These functions have only partly been superseded by the United Nations Global Compact and the Organisation for Economic Co-operation and Development (OECD) guidelines for multinational corporations.

A new look at competition policy

The exercise of buyer power across national boundaries highlights the major weakness of global regulation of competition. Current competition policy allows abuses of buyer power because:

- imperfect markets can defy standard economic analysis and provide a big challenge to regulators;
- power can be more of a reflection of size than monopoly;
- competition policy focuses primarily on consumer rather than producer welfare (Vorley, 2003).

Economic globalization has made it necessary to improve world governance on questions of monopoly and competition. No international competition standards exist to regulate corporate activity from one continent to another. In the UK, for example, the authorities' remit is over UK or European Union (EU) consumers to protect their welfare against monopoly and seller power, and does not extend to overseas producers. If a UK-based company exerts buyer power to push down producer prices when it or its suppliers buy cocoa in Ghana or beans in Kenya, this will be a matter for the Ghanaian or Kenyan competition authorities.

There is heated debate as to whether the World Trade Organization (WTO) is the right forum to address global competition issues. The development of a WTO Competition Law Framework is heading in a very different direction: simplifying regulation across national boundaries to facilitate transnational commerce and market access for industrialized country goods and services.

Fairness in trading and 'responsible supply chain management'

Stock markets reward buyer power, seeing it as a measure of a 'sustainable business' that will generate competitiveness, profits and shareholder value. Thus, voluntary self-regulation as a tool for improving agri-food companies' dealings with their suppliers and, ultimately, with small- and family-scale producers will be limited by shareholder pressure and company mindset. Public policy leverage over private-sector governance of food chains is weak and poorly defined. Equity and fairness in trading are almost entirely absent from the gamut of benchmarks, codes and standards for corporate social responsibility (CSR). Very few corporations seem to have made any significant moves to bring the CSR agenda onto their buying desks, the sharp end of trade with their supply chain. They remain resolutely customer, rather than supplier, focused. Supermarkets, for example, have shown much more interest in reacting quickly to technologies that alienate consumers (such as genetic modification) than in reacting to marketing practices that alienate suppliers. Price wars and pressure on suppliers and farmers are conducted in the name of providing customer value.

Retailers point to a commitment to Fairtrade labelled goods as a sign of commitment to trade justice. But Fairtrade labelling alone is a weak proxy for company commitment to fairness and justice in trading. This is due, first, to its limited market penetration; even in the UK, which has a relatively well-developed fair trade market, sales accounted for only 0.13 per cent of the UK£76 billion spent on food and drink in the UK in 2003, or 0.07 per cent if catering services are included. Second, many retailers have positioned fair trade as an upmarket de-commodified niche, rather than as a means of transforming their mainstream businesses. In effect, retailers have made fairness and justice in trading a consumer choice rather than a corporate standard.

Does fair trade have lessons for mainstream trading with small-scale primary producers and intermediaries? Fair trade has four key elements:

1 direct purchase;
2 guaranteed minimum price and price premiums;
3 credit allowances;
4 long-term relationships.

Incorporating these principles (or at least the last three) within contracts on a much wider scale can be implemented without being trumpeted as 'fair trade' or branded as a 'fair trade store'; rather, it becomes a corporate standard whereby customers walking into a store or buying a brand are reassured that their purchases have not contributed to the exploitation of producers and workers. A branded food manufacturer or retailer could take this further and apply fair trade concepts to all of its trade with the developing world.

Another ingredient of fair trade in non-plantation sectors is procurement from smallholders. A corporate commitment to rethinking supply chain management in favour of smallholders would be another step towards re-governing markets in support of poverty reduction. Of course, smallholder-based production does not necessarily represent the most 'sustainable' or 'ethical' options if we consider all issues, such as labour standards and worker representation. Inevitably, there will be some trade-offs between systematized 'ethics' via plantations and 'sustainable livelihoods' via smallholders.

Conclusions

It is important to evaluate the risks and benefits of strategies to connect small-scale producers with retail-driven chains before promoting risky, capital-intensive moves by small-scale producers into possibly rapidly saturated markets. The high requirements for entering buyer-driven chains, including investments in information and logistics, and compliance with 'voluntary' standards, codes and benchmarks, can profoundly affect farmers' access to (and entry into) markets. This means that the higher land and labour efficiency of smallholder production is no longer necessarily a comparative advantage. The connection between agriculture and poverty alleviation is thereby weakened. Given that co-operative structures and other forms of organization which allow smallholders to 'act big' do not appear to be the 'silver bullet' that many have hoped, finding other ways to re-govern agri-food markets at a more structural level in favour of smallholders remains a priority.

References

Bebbington, A. (2001) 'Globalized Andes? Livelihoods, landscapes and development', *Ecumene*, vol 8, no 4, pp414–436

Berdegué, J. A. (2001) *Cooperating to Compete: Associative Peasant Business Firms in Chile*, PhD thesis, Wageningen University, The Netherlands

Burt, S. L. and Sparks, L. (2003) 'Power and competition in the UK grocery market', *British Journal of Management*, vol 14, no 3, pp237–254

Fajarnes-Garces, P. and Matringe, O. (2002) *Recent Developments in International Banana Marketing Structures*, UNCTAD, Geneva

Gereffi, G. (1994) 'The organization of buyer-driven global commodity chains: How US retailers shape overseas production networks', in Gereffi, G. and Korzeniewicz, M. (eds) *Commodity Chains and Global Capitalism*, Praeger, Westport, Connecticut, pp95–122

Hernández R., Reardon, T. A., Berdegué, J. A., Balsevich, F. and Jano, P. (2004) 'Acceso de pequeños productores de tomate a los supermercados en Guatemala', www.regoverningmarkets.org

Mainville, D. and Reardon, T. (2004) 'Modelling relations among supermarkets, intermediaries and produce farmers in Brazil', Paper presented at EAAE Conference on Retailing and Producer–Retailer Relationships in the Food Chain, Paris, 5–6 May 2004

Morisset, J. (1997) *Unfair Trade? Empirical Evidence in World Commodity Markets Over the Past 25 Years*, World Bank–Country Economics Department, www.worldbank. org/html/dec/Publications/Workpapers/WPS1800series/wps1815/wps1815.pdf

Muñoz, D. with Cruz, B. and Canedo, M. (2004) *Organizaciones Económicas y Políticas Públicas: Un Estudio Comparativo,* Editiones Plural, La Paz

Östensson, O. (2002) *Commodities in International Trade: Current Trends and Policy Issues Implications for Caricom Countries*, UNCTAD, Geneva

Reardon, T. and Berdegué, J. A. (2002) 'The rapid rise of supermarkets in Latin America: Challenges and opportunities for development', *Development Policy Review*, vol 20, no 4, pp371–388

Spoor, M. (1997) *The 'Market Panacea': Agrarian Transformation in Developing Countries and Former Socialist Economies*, Intermediate Technology Publications, London

UK Competition Commission (2000) *Supermarkets: A Report on the Supply of Groceries from Multiple Stores in the United Kingdom,* Competition Commission, London, www.competition-commission.org.uk

Vorley, B. (2003) *Food, Inc: Corporate Concentration from Farm to Consumer,* UK Food Group, London

Wrigley, N. and Lowe, M. S. (2002) *Reading Retail: A Geographical Perspective on Retailing and Consumption Spaces*, Arnold, New York, and Oxford University Press, London

11

Concluding Reflections on the Future of Ethical Sourcing

Stephanie Barrientos and Catherine Dolan

The food system has undergone significant transformation during recent years. One dimension of this has been the rapid rise of ethical sourcing. Fair and ethical trade have become increasingly integrated within the mainstream of food production and retailing. There are many tensions between the commercial imperative for greater efficiency and lower cost that drives the global food system, and civil society pressure for fairer trade. A central question is the extent to which ethical sourcing is able to generate fairer trade and better conditions for the many small producers and insecure workers who provide the backbone to global food production. This volume has brought together a number of chapters on ethical sourcing in the global food system. These have explored some of its successes, but also the many challenges that ethical sourcing faces as it expands within the mainstream. Here we reflect on some of the key questions facing ethical sourcing as it advances in the future.

Opportunities for ethical sourcing

Food is now sourced from all parts of the world, creating wide choice and year-round availability at low prices. Large food producers and retailers play an increasingly dominant role, controlling a global food system driven by consumer demand. This has engendered a number of positive benefits, particularly for Northern consumers. They can now access food products in greater variety, convenience and quality, and relatively cheaper than ever before. However, as this volume has shown, enhanced consumer choice and year-round availability of cheap food has consequences for workers and small producers engaged in its production. During the past two decades, a growing lobby of civil society groups has begun to question the social impacts of global food production in both the North and South. These concerns have gained momentum through civil society campaigns and media publicity relating to unethical corporate behaviour. Many consumers are increasingly concerned about the social conditions under which their food is produced. A growing number now actively support fair and ethical trade, and are prepared to pay higher prices to ensure that the food they eat is produced and traded more fairly.

As consumer interest in ethical sourcing has grown, and civil society pressure becomes more intense, big retailers and brand manufacturers are recognizing the commercial value of an ethical policy and are positioning themselves as socially responsible companies. Many big corporate names now sell fair trade-labelled products or run fairly traded lines, and many operate codes of labour practice that cover workers employed in their supply chains. A plethora of initiatives have grown up around fair and ethical trade. In fair trade, these range from specialist alternative trade organizations (ATOs), through non-governmental organization (NGO) initiatives such as the Rainforest Alliance to Fairtrade certification under Fairtrade Labelling Organizations International (FLO). In ethical trade, numerous individual food retailers and producers now apply codes of labour practice in their supply chains, and many are involved with independent multi-stakeholder initiatives, such as the Ethical Trading Initiative (ETI) and Social Accountability International (SAI). Ethical sourcing is thus beginning to penetrate the mainstream of the global food system. But, as we have seen, this has also generated a number of challenges.

Challenges for ethical sourcing

Tensions between the commercial imperatives driving the global food system and ethical sourcing are not going to be easily resolved. Retailers and food manufacturers now enjoy an unprecedented concentration of power, but continue to face intense commercial competition in the global economy. The battle to enhance market share and shareholder returns drives them to take aggressive measures to reduce their costs and maximize their revenues. Those who fail to keep up are likely to see share prices fall and face a financial squeeze. These pressures can have adverse effects on those at the bottom of the food chain. Small producers find it harder to overcome the rising barriers to supplying global food markets, and many workers face increasingly insecure and informal employment as migrant and contract labour. The commercial pressures operating in the global food system can thus be at odds with the principles of ethical sourcing. For some companies, ethical sourcing fits with their corporate philosophy and philanthropic background; but for many it appears to run counter to a primary focus on maximizing efficiency and profit. These tensions give oxygen to those that criticize corporate engagement with fair and ethical trade as 'green washing'.

Debates over corporate commitment to ethical sourcing have become more prevalent with the adoption of fair trade lines by an increasing number of multinational food manufacturers. Some stakeholders within the fair trade movement, who have long perceived multinational corporations as an obstacle to trade equity, question whether fair trade should engage with large companies or retain a principled distance. Some ATOs are concerned that supporting multinational engagement in fair trade erodes the movement's founding principles and the long-term viability of fair trade products. They worry that corporate interest in the ethical market is opportunistic and lacks the commitment required to sustain producers through the boom-and-bust cycles of international trade. Some ATOs could find it increasingly difficult to survive since large multinationals provide a growing outlet for fair trade products. On the other hand, proponents of mainstreaming believe that a Fairtrade-labelled product, whether sold by Nestlé, Kraft or Ten Thousand Villages, undergoes exactly the same certification process. For them, multinational food companies offer an unprecedented opportunity to grow the fair trade market and extend its benefits to a wider swathe of small producers.

The entry of large supermarkets into the development of own-brand fair trade lines adds another dimension to this challenge. Under existing FLO rules, supermarket own-brand products may carry the Fairtrade logo, without the supermarket itself being licensed by Fairtrade Labelling Organizations International (FLO) so long as the products are sourced from the FLO register. Supermarkets are thus not formally required to adhere to the same rigorous fair trade standards as ATOs or even the large food manufacturers that package and label fair trade products. This raises questions as to whether supermarkets will extend their conventional purchasing practices to the sourcing of fair trade? Could they play producer groups off against each other and undermine stable supply relationships by switching between them? There have already been criticisms in the press that large retailers are taking advantage of ethical consumers' willingness to pay more for fair trade by levying a high mark up on the selling price (Stecklow and White, 2004). However, sales into supermarket own-brand lines still bring producers the benefits guaranteed by the FLO register and have facilitated the growth of a whole new range of fair trade products, particularly perishable items, such as fresh produce and flowers, which often require cool chain conditions to survive. This growth has played an important role in the expansion of fair trade sales, extending the opportunities provided by fair trade to a wider range of producers.

Increased growth also raises questions for the long-term position of small-scale producers in fair trade schemes. As fair trade expands and widens the scope of its product range, FLO is increasingly certifying 'plantations' and larger commercial farms. This, too, has created tensions within the fair trade movement. Some are concerned that certifying plantations has moved fair trade away from the original goals of the movement to support poor and marginalized producers and does little to alter pre-existing trade relationships. Others argue that the benefits of fair trade should also extend to plantation workers, whose numbers are increasing in an ever-more consolidated global food system. Is fair trade conducting an uphill battle on this front? Should the fair trade movement try to hold back the tide of growth, or should it be looking for more innovative solutions to support small-scale producers as they adapt to an increasingly concentrated global food environment?

In ethical trade, similar dilemmas prevail. Global food production is based on complex supply chains. The upper tiers are occupied by producers and packers, often in large-scale commercial enterprises, who typically employ a core of permanent or regular workers. Many of these workers might benefit from codes of labour practice. They may also be members of trade unions and covered by collective bargaining agreements. But below the first tier there is often a network of smaller producers and out-growers who experience much poorer conditions. At all tiers of production there is a growing need for an army of casual workers, often female, migrant and recruited by third-party contractors. Employment conditions for these groups are often poorest. Codes of labour practice appear less able to reach these more insecure workers, and they are least likely to be organized through traditional trade unions.

The issue of whether the purchasing practices of large producers and retailers generates the employment of a 'casualized' workforce and contributes to poor labour conditions is unlikely to abate. Civil society organizations have been able to stimulate the setting-up of ethical sourcing teams by large food companies; but, to date, they have had less effect on the commercial operations of their buying departments. Some question whether voluntary codes of labour practice can have any real effect. Global Union Federations such as the International Union of Foodworkers are putting increasing energy into the negotiation of international framework agreements with multinationals. Some NGOs, such as ActionAid, are focusing their campaigns on the need for regulation of large food companies, rather than voluntary initiatives (ActionAid, 2005). Whether these will be more effective in the context of a globalized economy remains to be seen. In the long run, it could be that a combination of strategies is required, both voluntary and regulatory. But the 'purchasing practices' issue highlights perpetual tensions between commercial imperatives and ethical principles in the global food system. In this context, pressure from trade unions and NGOs will have to persist to ensure that ethical sourcing does not slip under the commercial radar screen.

Emerging issues

As ethical sourcing continues to advance within the mainstream of the global food system, it is likely to continue to face issues that will test its viability. An important question is the extent to which it addresses poverty amongst the groups that are reached. A number of studies have been initiated to advance participatory monitoring and evaluation, and assess impact, both in fair and ethical trade (Ronchi, 2002; Barrientos and Smith, 2004 and 2006). These assessments are still at an early stage; but their findings are unlikely to be clear cut. In any assessment of poverty impact, it is difficult to isolate the effect of a specific intervention from wider influences affecting change. Many producers who are engaged in fair trade benefit from minimum prices and better supply relationships. But sales of fair trade are often only a small proportion of their total sales, and the benefits typically affect a relatively small proportion of producers. However, fair trade is not simply about price. It can also bring less tangible benefits through creating opportunities for capacity-building and better export market linkages. In ethical trade, case studies indicate that while regular and permanent workers benefit from codes, they are less effective in reaching more insecure and marginal workers. But ethical trade has moved the labour conditions of global workers up the agenda. It is part of a growing move to address international labour standards amongst an increasingly informalized global workforce. Ethical sourcing may therefore only be able to demonstrate that it is benefiting a limited number of producers and workers. But its wider effects are less easy to discern, and its longer-term role may be its active contribution to a broader process of change.

There are also opportunities to pursue new directions in ethical sourcing. The rapid rise in supermarkets in parts of the global South (Latin America, Africa and Asia) presents a new opportunity to expand fair trade consumption into the developing world. A number of Southern-based fair trade initiatives are emerging. In Mexico, for example, some civil society groups who recognized the limitations of a Northern fair trade market formed Comercio Justo México (Fair Trade Mexico) in 1999 to build a domestic fair trade market (Jaffee et al, 2004). In Latin America and Africa, there are also moves to coordinate different national initiatives and to expand fair trade regionally. Within ethical trade, a number of Southern initiatives have emerged to develop and implement local codes of labour practice. In this volume we have cited examples in South Africa, Kenya and Zimbabwe. These initiatives aim to bring together suppliers, trade unions and NGOs in a particular sector or country to develop a local code and monitoring systems. They face many challenges, particularly the difficulty of negotiating collaboration between groups who have often historically experienced tensions between participants. However, such initiatives have the potential of developing a more embedded and sustainable approach to ethical sourcing and redress a Northern bias that has often existed in these consumer-oriented schemes.

Southern initiatives provide the opportunity for greater local flexibility. But will they contribute to the plethora of different voluntary initiatives and labels that make up ethical sourcing? Both fair and ethical trade have produced an array of schemes. These can be confusing for consumers, producers and workers. The creation of FLO played an important role in bringing together the different national fair trade certification initiatives into one labelling organization in 1997. There have also been recent moves within ethical trade to bring together a range of different initiatives, including ETI and SAI, through the formation of a joint initiative in 2004. This is still at a pilot stage and it is too early to know how it will advance. Greater harmonization between diverse initiatives helps to reduce multiplicity of schemes. But it may be difficult to negotiate agreement in an environment where views on the future strategy of ethical sourcing are diverse.

As both fair and ethical trade expand, there is increasing overlap between the issues that they address. To date, there has been a clear demarcation – with fair trade focused on small producers and ethical trade on labour issues. But as fair trade engages more with plantations and larger commercial farms employing workers, and ethical trade engages more with purchasing practices along the supply chain, the overlap between them will increase. Given increasing inter-linkages between the agendas of fair and ethical trade, should there be greater collaboration and/or an integration of the two? To date, they have remained fairly distinctive. But they also jointly form part of a wider movement for fairer trade, which also enhances their strength.

It is important to acknowledge that fair and ethical trade are not a panacea for the deep-seated inequities that characterize the global economy, but rather two related strands within a wider movement for the reform of trade relations. Global trade in agriculture is now at a crucial turning point in terms of attempts to address global rules and trade policy that systematically disadvantage poorer countries. Many NGOs and global unions are engaged in campaigns that take a broader approach to trade reform. The Trade Justice Campaign, for example, has brought together a vast array of civil society actors, from trade unions and NGOs to women's institutes and churches, to campaigns for more equitable trade for developing countries. It is tackling fundamental issues in relation to the World Trade Organization (WTO). To date, international trade rules in agriculture have favoured the narrow commercial interests of the most powerful trading nations and the largest corporations. In order to rebalance the global trading system, international trade rules and institutions must be geared towards a more equitable system of international agreements aimed at sustainable development, poverty eradication and the promotion of human rights.

The rise of ethical sourcing demonstrates the success that NGOs, trade unions and consumer groups have had in forging an alternative vision of trade. This is a vision that provides opportunities for marginalized producers and workers to participate in the global food system on more favourable terms. It is encapsulated in FINE's declared goal of enhancing sustainable development 'by offering better trading conditions to, and securing the rights of, marginalized producers and workers – especially in the South'.[1] A decade ago, few would have dreamed that ethical sourcing could have advanced its position within the mainstream of the global food system to the extent that it has today. Voluntary approaches to ethical sourcing alone are likely to remain limited in their scope. Nevertheless, they constitute an important part of a wider movement to seek greater fairness in the trading system that is likely to continue to grow in future years. As it expands, however, ethical sourcing faces a number of challenging questions. These will not be easily resolved by debate alone, but are likely to be worked out in practice as transformation of the global food system advances.

Notes

1   FINE brings together the following umbrella organizations: Fairtrade Labelling
    Organizations International (FLO); the International Fair Trade Association
    (IFAT, formerly the International Federation of Alternative Trade); the Network of
    European World Shops (NEWS); and the European Fair Trade Association (EFTA).
    The FINE definition of fair trade was accessed from the British Association for
    Fair Trade Shops (BAFTS) website, www.bafts.org.uk/fair-trade/fine.htm, on 26
    August 2003.

References

ActionAid (2005) *Power Hungry: Six Reasons to Regulate Global Good Corporations*,
    ActionAid, London
Barrientos, S. and Smith, S. (2004 and 2006) *ETI Impact Assessment Report, Phase 1
    and 2*, Ethical Trading Initiative, London, www.eti.org.uk
Jaffee, D., Kloppenburg, J. and Monroy, S. (2004) 'Bringing the "moral charge" home:
    Fair trade within the North and within the South', *Rural Sociology*, vol 69, no 2,
    pp169–196
Ronchi, L. (2002) *Monitoring Impact of Fairtrade Initiatives: A Case Study of Kuapa
    Kokoo and the Day Chocolate Company*, Twin, London
Stecklow, S. and White, E. (2004) 'What price virtue? At some retailers, fair trade
    carries a very high cost', *The Wall Street Journal*, 8 June 2004

Glossary of Selected Fair and Ethical Trade Organizations

Alternative trade organizations (ATOs) and companies

AgroFair (The Netherlands): www.agrofair.com
AgroFair is an importer and distributor of Fairtrade and organic tropical fresh fruit. The company is co-owned by its producers and represents them in the European market. AgroFair markets bananas under its own brands: Oké (Fairtrade) and Eko-Oké (organic Fairtrade).

Altertrade Japan: www.altertrade.co.jp/english/atjhp1.htm
Altertrade is engaged in a unique 'people-to-people' trade in the developing world, handling 'eco-food' products. The organization started from helping the Negros islanders of the Philippines who were reeling after a catastrophic sugar market collapse during the early 1980s.

Body Shop Community Trade Programme: www.thebodyshop international.com/web/tbsgl/values_sct_what.jsp
The Body Shop Community Trade programme is aimed at small producer communities around the world who supply the Body Shop with accessories and natural ingredients. It ensures a fair deal for the producers and their communities, enabling them to have more control over their futures.

Equal Exchange (US): www.equalexchange.com
A worker-owned co-operative, Equal Exchange offers consumers fairly traded gourmet coffee direct from small-scale farmer co-ops in Latin America, Africa and Asia.

EZA 3.Welt (Austria): www.eza3welt.at/eindex.htm
A fair trade organization dealing with marginalized producer groups from developing world countries. It was founded in 1975 and since then has been by far the most important fair trader in Austria. EZA 3.Welt is the Austrian member of the European Fair Trade Association.

Fair Trade Online (UK): www.store.yahoo.com/fairtradeonline-uk
Online fair trade store in conjunction with Oxfam and Traidcraft.

Fair Trade Organization (The Netherlands): www.fairtrade.nl
The Fair Trade Organization handles an assortment of 3000 products purchased from craftsmen and farmers in Africa, Asia and Latin America for Dutch and Belgian markets. Part of the organization is Fair Trade Assistance, a consultancy designed to assist fair trade producers with accessing the European export market.

gepa Fair Handelshaus (Germany): www.gepa3.de/index.html
Gepa is Europe's biggest fair trade company, with an annual turnover of more than 33 million euros. It transfers over 18 million euros for food products, handicrafts and textiles to its business partners in over 150 co-operatives and marketing organizations in Africa, Asia and Latin America.

NovoTRADE (The Netherlands): www.novotrade.nl
NovoTRADE is a consultancy set up in alliance with Twin Trading in the UK. Twin and NovoTRADE facilitate local support structures and activities, providing resources and backstopping and linking them to a wider network of traders, banks and support agencies. Through mutual support between producer organizations, the impact is extended to a wider group of producers on a cost-effective basis.

Oxfam (UK): www.oxfam.org.uk/what_we_do/fairtrade/ft_food.htm
Oxfam retails fair trade goods through its network of over 300 charity shops in the UK and Ireland. Oxfam used to source Fairtrade but ceased to do so in 2003/2004. Oxfam shops continue to provide an outlet for other Fairtrade brands.

Peoplink (US): www.peoplink.org
A non-profit online marketplace that enables consumers to buy artisan goods directly from artisans all over the world. It also contains information about fair trade partners and suppliers in the South.

SERRV International (US): www.serrv.org
A non-profit alternative trade and development organization. SERRV has been working to assist artisans and farmers for more than 55 years and is an accredited member of the International Fair Trade Association. SERRV sells fair trade products through its online store.

Trade Aid (New Zealand): www.tradeaid.org.nz
A fair trade organization established in 1973, selling fair trade products and food through its network of shops. The organization is also involved in campaigning and information dissemination.

Traidcraft (UK): www.traidcraft.org.uk
An active community of supporters, shareholders, customers, professionals and producers committed to fighting poverty through trade. As the UK's leading fair trade organization, Traidcraft plc tackles poverty by providing a market for more than 100 fair trade producer groups to supply. Traidcraft plc's sales are now worth more than UK£12 million a year, providing vital income for producers in over 30 countries.

Twin (UK): www.twin.org.uk
Established in 1985, Twin is a leading alternative trading company in the UK.
Twin is committed to:

- developing long-term trading relationships built on trust with producers;
- bringing the producer and the market closer together;
- increasing the ability of small producers to trade on an equal footing;
- avoiding a dependency on Twin;
- the participation of producers in Twin decision-making processes.

Solidar'Monde (France): www.solidarmonde.fr
A French fair trade importer established in 1984 that imports craft products
and foodstuffs from developing world countries and markets them in France.
These products come from about 120 organizations in 40 countries in Latin
America, Africa and Asia, and are supplied to several hundred specialized
organic and fair trade shops in France.

Labelling initiatives

**Fairtrade Labelling Organizations International (FLO):
www.fairtrade.net**
FLO is the worldwide Fairtrade standard-setting and certification organization.
It regularly inspects and certifies approximately 420 producer organizations in
50 countries in Africa, Asia and Latin America, permitting more than 800,000
producers, workers and their dependants to benefit from labelled Fairtrade.
FLO guarantees that products sold anywhere in the world with a Fairtrade
label marketed by a national initiative conform to Fairtrade standards and
contribute to the development of disadvantaged producers. FLO has 20
national initiatives:

- Australia/New Zealand: Fairtrade labelling (www.fta.org.au);
- Austria: Fairtrade Austria (www.fairtrade.at);
- Belgium: Max Havelaar Belgium (www.maxhavelaar.com);
- Canada: Transfair Canada (www.transfair.ca);
- Denmark: Max Havelaar Denmark (www.maxhavelaar.dk);
- France: Max Havelaar France (www.maxhavelaarfrance.org);
- Germany: TransFair Germany (www.transfair.org);
- Italy: TransFair Italy (www.transfair.it);
- Ireland: Fairtrade Mark Ireland (www.fair-mark.org);
- Japan: Fairtrade label Japan (www.fairtrade-jp.org);
- Luxembourg: TransFair Minka (www.transfair.lu);
- Mexico: Comercio Justo México (www.comerciojusto.com.mx);
- The Netherlands: Stichting Max Havelaar (www.maxhavelaar.nl);
- Norway: Max Havelaar Norge (www.maxhavelaar.no);
- Finland: Reilun kaupan edistämisyhdistys.ry (www.reilukauppa.fi);

- Spain: Asociación para el Sello de Comercio Justo España (www.sello comerciojusto)
- Sweden: Föreningen för Rättvisemärkt (www.raettvist.se);
- Switzerland: Max Havelaar Stiftung Schweiz (www.maxhavelaar.ch);
- UK: Fairtrade Foundation (www.fairtrade.org.uk);
- US: Transfair USA (www.transfairusa.org).

FLO comprises two organizations:

1 FLO e. V. is a multi-stakeholder association involving FLO's 20 national initiatives, producer organizations and traders that develops and reviews standards and assists producer groups to gain and maintain certification.
2 FLO-Cert coordinates all tasks related to the inspection of producers, trade auditing and certification.

Sample fair trade brands

Cafédirect: www.cafedirect.co.uk
Cafédirect is the UK's largest Fairtrade hot drinks company. Its brands, Cafédirect, 5065, Teadirect and Cocodirect, are sold through most of the major supermarkets. It buys from 33 producer organizations in 11 countries.

Clipper Teas: www.clipper-teas.com/is_fairtrade.htm
Clipper was the first tea company in the world to have been awarded the Fairtrade Mark (1994) and has since become established as the leading brand in Fairtrade teas, with an extensive range of high-quality Single Estate Fairtrade teas, as well as its own unique blend, Clipper Fairtrade Tea, which is available in all major UK supermarkets.

Divine Chocolate: www.divinechocolate.com
Fairly traded chocolate made from cocoa sourced from small-scale farmers that are part of the Kuapa Kokoo cocoa farmers co-operative in Ghana. Divine chocolate is supplied by The Day Chocolate Company, which is part-owned by Kuapa Kokoo, Twin Trading and Body Shop International.

Fair trade advocacy and information organizations

Banana Link (UK): www.bananalink.org.uk
Banana Link aims to alleviate poverty and prevent further environmental degradation in banana-exporting communities and to work towards a sustainable banana economy. It aims to achieve this by working cooperatively with partners in Latin America, the Caribbean, West Africa and the Philippines, and with a network of European and North American organizations.

Café Unidos (Canada): www.cafeunidos.org
Café Unidos is a website designed to provide information on sustainable coffee and to foster relationships between coffee growers, sellers and buyers. The site is

a joint collaboration of Equiterre and the Eco-Research Chair in Environmental Law and Policy at the University of Victoria, British Columbia.

Équiterre (Canada): www.equiterre.qc.ca

A not-for-profit organization dedicated to promoting ecological and socially just choices through action, education and research from a standpoint that embraces social justice, economic solidarity and the defence of the environment. Équiterre's Fair Trade Programme works towards the creation of a more equitable system of international trade through the promotion of fair trade and other consumer alternatives. Équiterre's programme includes an educational campaign directed at both consumers and businesses called A Just Coffee. The programme is also doing research towards the development of other fair trade products and economic alternatives.

Ethical Junction (US): www.ethical-junction.org

Ethical Junction is a one-stop shop for ethical organizations and ethical trading. The site has an ethical shopping centre and a search engine that only searches ethical organizations, and the directory is full of hundreds of organizations representing a huge range of businesses and interests. From ethical financial advice, fairly traded goods and solar panels to business consultants, theatre groups and campaign organisations, Ethical Junction is the place for shopping and information. The Ethical Junction was formed to make it easier for us all to adopt a more ethical and rewarding lifestyle, and to contribute to making the world of commerce fair and sustainable.

Fair Trade Resource Network (US): www.fairtraderesource.org

The Fair Trade Resource Network raises consumer awareness about improving people's lives through fair trade alternatives.

Fair Trade Toronto (Canada): www.fairtradetoronto.com

Fair Trade Toronto's goal is to encourage Canadians to become informed and concerned about the conditions under which the products they consume have been developed or manufactured. Working in cooperation with communities, businesses, farmers, NGOs and governmental organizations, they aim to develop public awareness about the fair trade alternative, and to respond to consumer demands for fairly traded products.

Get Ethical (US): www.getethical.com

Online source for ethical products and ethical matters, this magazine is for the socially aware reader.

Global Exchange (US): www.globalexchange.org/campaigns/fairtrade

Global Exchange is an international human rights organization dedicated to promoting environmental, political and social justice. Since its founding in 1988, it has increased the US public's global awareness, while building partnerships worldwide.

**Organic Consumers' Association (OCA) (US):
www.organicconsumers.org**
Campaigning for food safety, organic agriculture, fair trade and sustainability, the OCA promotes food safety, organic farming and sustainable agricultural practices in the US and internationally. It provides consumers with factual information that they can use to make informed food choices. Its campaign strategies include public education, activist networking, boycotts and protests, grassroots lobbying, media and public relations, and litigation. It publishes one print newsletter (*Organic View*) and one electronic newsletter (*Organic Bytes*).

People for Fair Trade (Australia): www.fairtrade.asn.au/index.htm
A voluntary network of people in Australia who are committed to fair trade with producers of goods through the support of education and alternative trade.

Shared Interest (UK): www.shared-interest.com
Shared Interest is a co-operative lending society that aims to reduce poverty in the world by providing fair and just financial services. It was started in 1990 and has some 8000 members who have invested more than UK£20 million. It uses the pooled savings of its members to facilitate fair trade.

Social Venture Network (US): www.svn.org
A non-profit network committed to building a just and sustainable world through business.

**Solidaridad (The Netherlands): www.solidaridad.nl/indexengels.
html**
A development organization for Latin America. Solidaridad has three programmes: Sustainable Economy and Fair Trade; Society Building and Human Rights; and Pastorate, Ethics and Culture.

United Students for Fair Trade (USFT) (US): www.usft.org
The United Students for Fair Trade is a national network of student organizations advocating fair trade products, policies and principles. The core objective of USFT is to raise the awareness of, and expand the demand for, fair trade alternatives, both on campuses and in communities.

Regional networks and trade networks

European Fair Trade Association (EFTA): www.eftafairtrade.org
The European Fair Trade Association is a network of 11 fair trade organizations in 9 European countries that import fair trade products from some 400 economically disadvantaged producer groups in Africa, Asia and Latin America. EFTA's members are based in Austria, Belgium, France, Germany and Italy.

Fair Trade Federation (FTF): www.fairtradefederation.com
The Fair Trade Federation is an association of fair trade wholesalers, retailers and producers, mostly from North America. FTF also acts as a clearinghouse for information on fair trade and provides resources and networking opportunities for its members.

Fairtrade Labelling Organizations International (FLO) Producer Support Office Ecuador/Peru: www.comerciojustoecuadorperu.net
The regional office for FLO in Ecuador and Peru is a unit of the Producer Support Network (PSN) of FLO. It cooperates in strengthening local fair trade organizations and in answering the demands of producer organizations.

Fairtrade Labelling Organizations International (FLO) Producer Support in Central America: www.flocentroamerica.net
The regional office of FLO for Central America was created in 1999 in order to respond better to the Fairtrade-related development needs of producer partners, as well as to emergency situations that producer organizations face. All aspects of the inspection of (new) organizations are undertaken by local inspectors and are directly coordinated by the Certification Unit of FLO in Bonn, Germany. The countries covered are Guatemala, El Salvador, Honduras, Nicaragua and Costa Rica.

International Fair Trade Association (IFAT): www.ifat.org
The global association of nearly 300 fair trade organizations in over 60 countries, IFAT brings its members together in order to develop the market for fair trade, build trust in fair trade and speak out for fair trade.

Network of European Worldshops (NEWS!): www.worldshops.org
NEWS! is the umbrella network of 15 national associations of worldshops in 13 different countries throughout Europe. Through the national associations, NEWS! represents the worldshops – shops that offer fairly traded products and campaign for a fairer world trade system. The national worldshop associations are:

- Austria: Arge Weltläden (www.weltlaeden.at);
- Belgium: Oxfam Wereldwinkels (www.oww.be);
- Denmark: Fair Trade Danmark (www.u-landsimporten.dk);
- Finland: Maailmankauppojen liitto (www.maailmankaupat.fi);
- France: Artisans du Monde (www.artisansdumonde.org);
- Germany: Weltladen-Dachverband (www.weltlaeden.de);
- Italy: Associazione Botteghe del Mondo (www.assobdm.it);
- Italy: CTM Altromercato (www.altromercato.it);
- The Netherlands: Landelijke Vereniging van Wereldwinkels (www.wereldwinkels.nl);
- Portugal: Coordenação Portuguesa de Comércio Justo (www.modevida.com/comercio2.html)

- Spain: Coordinadora Estatal de Commercio Justo (www.e-comerciojusto. org);
- Sweden: Världsbutikerna för Rättvis Handel (www.varldsbutikerna.org);
- Switzerland: Claro Fair Trade plc (www.claro.ch/index.html);
- Switzerland: Association Romande des Magasins du Monde (www.mdm. ch);
- UK: British Association for Fair Trade Shops (BAFTS) (www.bafts.org. uk).

Southern fair trade partners (examples)

AMKA (Tanzania): www.catgen.net/amka/EN/0.html
A Tanzanian NGO set up in 1994 to provide business development services and export market facilitation to small and medium enterprises in Tanzania. AMKA specializes in fair trade export.

Kuapa Kokoo (Ghana): www.kuapakokoo.com
Established in 1993, Kuapa Kokoo is a cocoa farmers' co-operative organization that works to improve the lot of their members. Kuapa Kokoo is a composite organization with five main subgroups (Kuapa Kokoo Farmers' Union; Kuapa Kokoo Limited; Kuapa Kokoo Farmers' Trust; Kuapa Kokoo Credit Union; and Day Chocolate Company – London).

Thandi (South Africa): www.thandi.com
Thandi's aim is to empower previously disadvantaged farming communities. With support and mentorship from leading players in the fruit and wine industries, these communities export top-class produce to countries all over the world.

Vinculación y Comunicación Social (Mexico): www.vinculando.org/ en/index.html
Vinculación y Comunicación Social is a non-profit association based in Mexico City, whose main goal is to support small producers and indigenous people to find appropriate commercial partners to sell their goods, both sustainable and conventional.

Multi-stakeholder social codes of practice (general)

Ethical Trading Initiative (ETI): www.ethicaltrade.org
The Ethical Trading Initiative is an alliance of companies, NGOs and trade union organizations. It exists to promote and improve the implementation of corporate codes of practice that cover supply chain working conditions. Its ultimate goal is to ensure that the working conditions of workers producing for the UK market meet or exceed international labour standards.

Global Reporting Initiative (GRI): www.globalreporting.org
The Global Reporting Initiative is a multi-stakeholder process and independent institution whose mission is to develop and disseminate globally applicable sustainability reporting guidelines.

Social Accountability International (SAI): www.cepaa.org
Social Accountability International works to improve workplaces and combat sweatshops through the expansion and further development of the international workplace standard, SA 8000, and the associated SA 8000 verification system.

Inter-governmental codes of practice

International Labour Organization (ILO) Declaration on Fundamental Principles and Rights at Work: www.ilo.org/public/english/standards
Adopted by the International Labour Conference in 1998, this declaration renews commitment among members to respect basic worker rights, such as the right to bargain collectively and pledges to end such practices as compulsory labour and child labour. A follow-up to 1977's Tripartite Declaration of Principles Concerning Multinational Enterprises and Social Policy.

ILO Tripartite Declaration of Principles Concerning Multinational Enterprise and Social Policy: www.ilo.org/public/english/standards/ norm/sources/added
Adopted in 1977 by the Governing Body of the International Labour Organization following similar declarations of the Organisation for Economic Co-operation and Development (OECD) and the United Nations Commission on Transnational Corporations.

OECD Guidelines for Multinational Enterprises: www.oecd.org
Adopted by the member countries of the OECD in June 2000, these guidelines cover fair trade issues ranging from employment relations to competition, taxation and bribery.

OECD
Principles for Corporate Governance: www.oecd.org
This is the OECD treatise for shareholder rights and other governance issues.

United Nations Global Compact: www.unglobalcompact.org
This compact comprises nine United Nations principles for multinational and transnational corporations. It recognizes various forms of human rights, labour/worker rights and environmental rights.

Social codes of practice: Agriculture

International Flower Code: International Union of Food, Agricultural, Hotel, Restaurant, Catering, Tobacco and Allied Workers' Associations (IUF): www.iuf.org
The German section of the international human rights organization FoodFirst Information and Action Network (FIAN) and the Friedrich-Ebert-Stiftung (FES), in collaboration with a number of other NGOs. Its model code incorporates core International Labour Organization (ILO) conventions and covers employment conditions, and health and safety for flower workers.

Liaison Committee Europe–Africa, Caribbean, Pacific (COLE–ACP) Harmonized Framework
COLE–ACP is the association of exporters, importers and other stakeholders in European Union ACP countries. The Harmonized Framework contains a set of core criteria that cover environmental conservation and social responsibility, including social welfare and workers' safety, as well as food safety, food hygiene and traceability.

Social codes of practice: Food

Banana Link (UK): www.bananalink.org.uk
Banana Link aims to alleviate poverty and prevent further environmental degradation in banana-exporting communities, and to work towards a sustainable banana economy. It aims to achieve this by working cooperatively with partners in Latin America, the Caribbean, West Africa and the Philippines, and with a network of European and North American organizations.

Ethical Tea Partnership (formerly Tea Sourcing Partnership) (UK): www.ethicalteapartnership.org
The Ethical Tea Partnership, formed in 1997, is an international organization with members throughout Europe and North America, which currently comprises the largest brands of tea in Western Europe. Over 47 tea brands sold in over 30 countries are now covered by the initiative. The Ethical Tea Partnership initiative aims to demonstrate that 100 per cent of the tea that their members buy is responsibly produced. Two elements are fundamental to the initiative: inclusivity and independent monitoring. Members of the partnership focus on six key areas of tea estates: employment (including minimum age and wage levels); education; maternity; health and safety; housing; and some basic rights.

Harkin-Engel Protocol: www.senate.gov/~harkin/specials/20011001-chocolate.cfm
The Harkin-Engel Protocol of October 2001 sets a four-year timetable for all stages of the cocoa industry to comply with standards set by the International

Labour Organization's (ILO's) Convention Against the Worst Forms of Child Labour, No 182. It has been signed by leading representatives of the US cocoa and chocolate industry and witnessed by the International Programme on the Elimination of Child Labour (IPEC) of the ILO, the International Union of Food and Allied Workers, the Child Labour Coalition, the National Consumers League and Free the Slaves.

International Cocoa Initiative: Working Towards Responsible Labour Standards for Cocoa Growing: www.chocolateandcocoa.org/News/press_release_070102.htm

The global chocolate and cocoa industry, in partnership with organized labour unions and NGOs, has established the International Cocoa Initiative to eliminate abusive child labour practices in cocoa cultivation and processing.

International Cocoa Organization: www.icco.org

Resolution on agricultural working practices. The international Cocoa Organization (ICCO) was established in 1973 to administer the first and subsequent International Cocoa Agreements. The 2001 Agreement, negotiated at the UN Cocoa Conference, puts emphasis on achieving a sustainable cocoa economy.

International Social and Environmental Accreditation Labelling (ISEAL): www.isealalliance.org

ISEAL members are collaborating in order to gain international recognition and legitimacy for their programmes; to improve the quality and professionalism of their respective organizations; and to defend the common interests of international accreditation organizations, while demonstrating the openness and transparency of operations that ISEAL members believe are fundamental to their integrity. The standards, certification programmes and accreditation systems developed by ISEAL members reflect a worldwide concern for social and environmental issues.

Rainforest Alliance Agriculture Programme: www.rainforest-alliance. org/programs/agriculture/index.html

The mission of the Rainforest Alliance's Sustainable Agriculture Programme is to integrate productive agriculture, biodiversity conservation and human development. Farmers, companies, co-operatives and landowners who participate in the programme meet comprehensive and rigorous social and environmental standards.

Social Accountability in Sustainable Agriculture (SASA): www.iseal alliance.org/sasa

The Sustainable Agriculture Network, Fairtrade Labelling Organizations International, International Federation of Organic Agriculture Movements and Social Accountability International have come together to investigate how best to ensure broad-based social accountability in agriculture.

UK Banana Group: www.bananalink.org.uk/future/alt_f.htm
In November 1998, the UK Banana Group, which includes all of the major banana-importing companies, produced a code of practice. The Banana Group Code has been largely superseded by the Ethical Trading Initiative's Base Code.

Wine Industry Ethical Trade Association (WIETA) (South Africa): www.wieta.org.za
WIETA is a not-for-profit voluntary association of many different stakeholders in the South African wine industry who are committed to the promotion of ethical trade in this sector, including wine producers, retailers, trade unions, NGOs and government.

Other relevant information

Make Trade Fair Campaign: www.maketradefair.org/en/index.htm
Oxfam International campaign involving a world-wide petition, coordination of events, research, information and briefings calling for decision makers to make trade fair.

***New Consumer*: www.newconsumer.org**
International fair trade magazine and organization covering all aspects of fair trade.

New Economics Foundation (NEF): www.neweconomics.org
Independent think-and-do tank that inspires and demonstrates real economic well-being. It aims to improve quality of life by promoting innovative solutions that challenge mainstream thinking on economic, environment and social issues.

The Institute of Development Studies: www.ids.ac.uk
A leading organization for research, teaching and communications on international development. It carries out research on workers and vulnerability, ethical trade and fair trade.

Trade Justice Movement: www.tradejusticemovement.org.uk
The Trade Justice Movement is a group of more than 70 organizations including trade unions, aid agencies, environment and human rights campaigns, fair trade organizations, faith and consumer groups. Together, they are campaigning for trade justice – not free trade – with the rules weighted to benefit poor people and the environment.

Index